THE LOVE OF
MONKEYS
AND APES

THE LOVE OF
MONKEYS
AND APES

DAN FREEMAN

octopus

CONTENTS

First published 1977 by Octopus Books Limited
59 Grosvenor Street, London W1

© 1977 Octopus Books Limited.   ISBN 0 7064 0612 5

Produced by Mandarin Publishers Limited
22a Westlands Road, Quarry Bay, Hong Kong

Printed in Hong Kong

INTRODUCTION
THE EVOLUTION OF MONKEYS

The true monkeys and apes as we know them form only a few end links in a chain of closely related animals. Collectively they are called primates and they range from the primitive lemurs, which have been isolated on the island of Madagascar off the east coast of Africa for more than 30 million years, to Man himself — a highly successful species which has colonized much of the land mass of the earth during no more than the last one million years. Before we scale, chapter by chapter, the rungs of the ladder that lead finally to Man, we must undertake a brief evolutionary survey of this delightful group as a whole. Over the years, some groups will have changed more or less than others; some will have retained characteristics they developed millions of years ago while others will have lost them. The detailed modifications of each family will emerge as we examine them more closely in the pages that follow this introduction. It is a fascinating story and to do it justice we must begin at the very beginning.

More than 70 million years ago, the mammals that occupied the earth were very similar. They were rather small, lived on the ground and fed mostly on insects. Their spread at this time may well have been inhibited by the presence of giant reptiles, for this was still the great age of the dinosaurs — a time when swamps, lakes and vast seas covered the earth and the only vegetation was ferns, conifers and primitive flowering plants.

But about 65–70 million years ago there was a dramatic change in the world's climate: a major shift from wet to dry. With this change the dinosaurs, so large and so cumbersome on land and yet so superbly adapted to the waters that had been their haven for more than 140 million years, foundered and perished in the muddy wastes that set hard under the desiccating wind of change — entombing them for ever.

As the climatic conditions changed, the flowering plants spread and gave rise to vast forests. This flowering boom was matched by an 'explosion' of insects which in turn were followed by their ravenous predators, the mammals and the birds. By 65 million years ago, during the Palaeocene era, competition between the mammals must have been so intense that some of them were forced to adopt a new life-style to survive. They took to the trees and in so doing found not only an untapped source of food to add to their insect diet, but also protection from ground

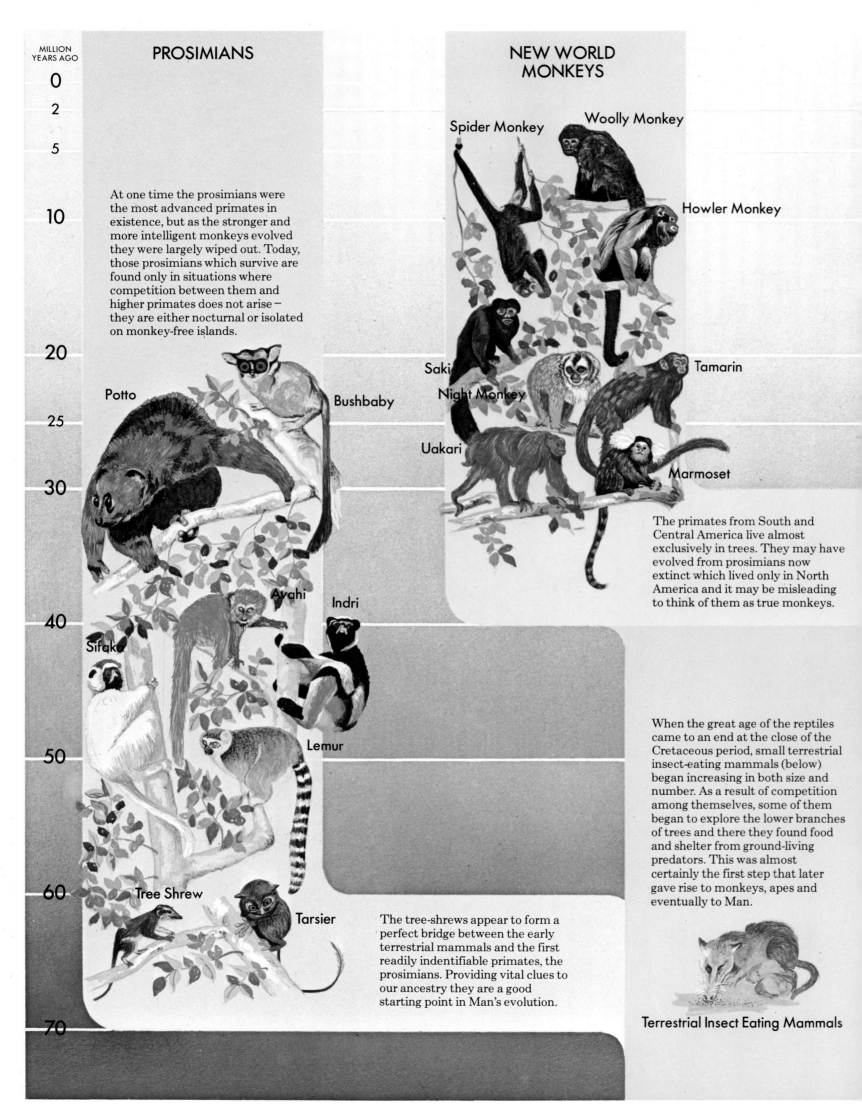

MILLION
YEARS AGO

0
2
5
10
20
25
30
40
50
60
70

PROSIMIANS

At one time the prosimians were the most advanced primates in existence, but as the stronger and more intelligent monkeys evolved they were largely wiped out. Today, those prosimians which survive are found only in situations where competition between them and higher primates does not arise – they are either nocturnal or isolated on monkey-free islands.

Potto

Bushbaby

Avahi

Indri

Sifaka

Lemur

Tree Shrew

Tarsier

The tree-shrews appear to form a perfect bridge between the early terrestrial mammals and the first readily indentifiable primates, the prosimians. Providing vital clues to our ancestry they are a good starting point in Man's evolution.

NEW WORLD MONKEYS

Spider Monkey

Woolly Monkey

Howler Monkey

Saki

Tamarin

Night Monkey

Uakari

Marmoset

The primates from South and Central America live almost exclusively in trees. They may have evolved from prosimians now extinct which lived only in North America and it may be misleading to think of them as true monkeys.

When the great age of the reptiles came to an end at the close of the Cretaceous period, small terrestrial insect-eating mammals (below) began increasing in both size and number. As a result of competition among themselves, some of them began to explore the lower branches of trees and there they found food and shelter from ground-living predators. This was almost certainly the first step that later gave rise to monkeys, apes and eventually to Man.

Terrestrial Insect Eating Mammals

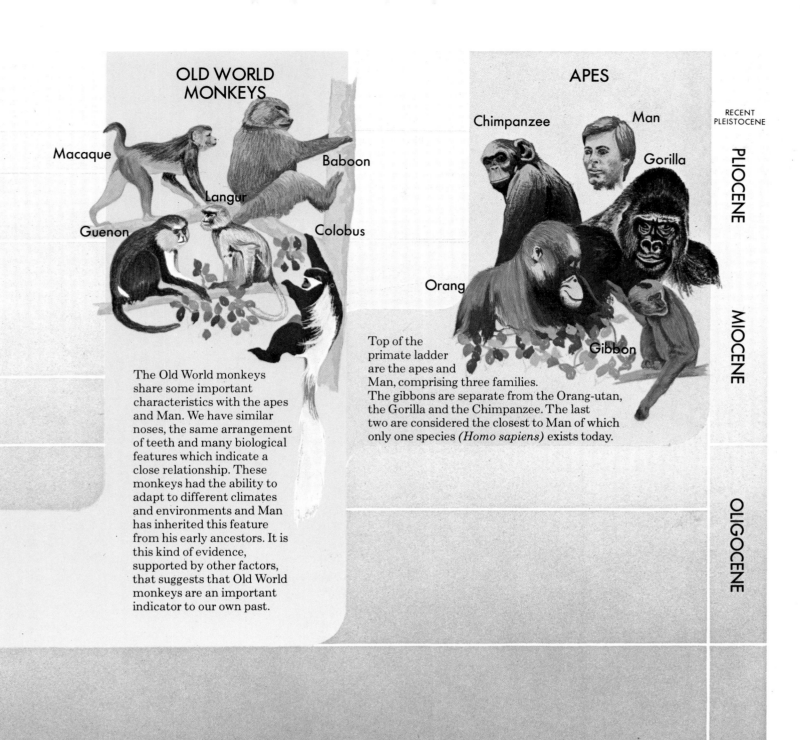

OLD WORLD MONKEYS

Macaque

Langur

Guenon

Baboon

Colobus

APES

Chimpanzee

Man

Gorilla

Orang

Gibbon

RECENT
PLEISTOCENE

PLIOCENE

MIOCENE

OLIGOCENE

EOCENE

PALAEOCENE

The Old World monkeys share some important characteristics with the apes and Man. We have similar noses, the same arrangement of teeth and many biological features which indicate a close relationship. These monkeys had the ability to adapt to different climates and environments and Man has inherited this feature from his early ancestors. It is this kind of evidence, supported by other factors, that suggests that Old World monkeys are an important indicator to our own past.

Top of the primate ladder are the apes and Man, comprising three families. The gibbons are separate from the Orang-utan, the Gorilla and the Chimpanzee. The last two are considered the closest to Man of which only one species *(Homo sapiens)* exists today.

This artist's impression of the evolution of the primates attempts to clarify a highly intricate process lasting millions of years. One way to understand evolution is to take all the living examples of a related group and to work backwards through time until a satisfactory common ancestor can be agreed upon. This is not always easy because scientists must rely upon available fossil evidence (which may still be incomplete), must be aware of the 'traps' set by parallel evolution (in which unrelated groups may grow to look the same as a result of similar environmental pressures) and must also agree which characteristics are the important ones on which to concentrate. Man and tree-shrews may represent the present-day extremes of a very complex process – even though nobody, at a glance, would ever guess that they had anything in common at all.

predators. The first primate-like creatures had come forward and identified themselves!

Among the earliest tree-dwelling mammals was a small tree shrew which still survives in the low growth of the forests of South-East Asia. Many zoologists are agreed that it is not actually a primate because it does not show the specific adaptations common to the group as a whole. But it is, nevertheless, important because it gives us somewhere to begin. It made the successful move, quite literally, of getting off the ground, and it persists today as evidence of what the very earliest tree-living creatures must have looked like.

From this starting block came the 'lower' primates, a group often referred to as the prosimians (which means 'nearly monkeys'). During the 20-million-year Eocene period the prosimians dominated forests in North America, Asia and Europe. During this time they underwent major physical changes. Their eyes gradually came to face forwards to give them binocular vision, enabling them to judge distances and to focus on objects in front of them without turning their heads to one side. Their noses flattened and became smaller to make room for these large eyes which were also taking over from long, sensitive nostrils as their most important sense organ. Up in the trees they largely lost the ability to walk on all four legs and instead became expert at jumping from branch to branch and from tree to tree, developing powerful leg muscles as they went. Their hands and feet were also changing. Nails replaced claws, leaving room for sensitive pads to develop on the tips of fingers and toes, and the thumb and big toe became 'opposable' — set apart from the other four digits, and indispensable for grasping food and for securing a safe grip on perilous treetop branches.

The hitherto unchallenged reign of the prosimians came to an end some 45–35 million years ago when much stronger kinds of primate began to evolve. Today these are known as the 'higher' primates, or anthropoids (which means 'human-like'), and they include all monkeys, apes and Man. The anthropoids were bigger, stronger and more intelligent than their fragile prosimian relatives and if there was any competition between them, the anthropoids invariably won. The prosimians, beaten into retreat, were forced to adapt to a nocturnal way of life in order to survive. Some of them also had the good fortune of being isolated on islands and other remote places which the anthropoids had not been able to occupy. But by 30 million years ago the anthropoids had completely usurped the prosimians as the dominant tree-dwelling primates.

Among these first anthropoids were the true monkeys — the monkeys which provide us with so much entertainment in zoos the world over as they leap across their cages at incredible speeds or swing nonchalantly by their strong 'prehensile' tails. These tails evolved only among the New World monkeys from Central and South America, giving them an extra limb for grasping branches. Virtually all the monkeys, however, developed loud chattering calls and piercing screams to keep in touch with each other in the dark forests where they travelled so extensively.

In their strength and ability to adapt and to travel, the monkeys spread out across the world, colonizing the great forests of the Americas, Europe, Africa and Asia. But about 35 million years ago, during the Oligocene period, the world's climate underwent yet another change. The warm, wet tropical luxuriance was gradually replaced by a prolonged period of cold and dry, and the great forests shrank back from the north and south until they hugged only the equatorial band between the tropics of Cancer and Capricorn. As the forests receded they carried the monkeys with them, forcing some into extinction. But they also left in their wake a few species, notably the hardy baboons and the macaques, to forsake the protection of the trees and to adopt a more terrestrial way of life.

By 30 million years ago the monkeys in the trees were living under intense competition for food and space — suffering the same pressures that had sent their distant ancestors clambering up those same tree trunks millions of years beforehand — and now there was a gradual trend towards returning to the ground. Some monkeys reached a compromise between the two ways of life, but within 5–10 million years the tailless apes were roaming the forest floors and learning how to stand upright although essentially they lived on all fours.

And so, within only the past few million years, during the late Pliocene era, Man has arrived on the scene to dominate the world — Man with his superior brainpower, intricate social structures and incredible technological achieve-

The Common Tree Shrew spends most of its life on or close to the ground. It normally leads a solitary life as it hurries over logs or turns over fallen leaves, searching every nook and cranny for insects. Occasionally more than one may be found together, but they will separate if disturbed and either make off across the forest floor or climb expertly into the low growth of bushes. In the bushes they also find fruit on which to feed, but insects probably form their staple diet and they are plentiful down on the ground. The tree shrews will sit on their haunches and hold their food between their forepaws, for they lack the ability to grasp food in one hand.
The females are usually pregnant for 46–50 days and the litter of two, or sometimes one, is the smallest of all the tree shrew litters. The young are weaned after about six weeks, at which point they leave the safety of their nest and fend for themselves in the inhospitable forests.

ments, all of which have enabled him to create his own environment and to virtually reshape the world as he sees fit.

The irony, of course, is that we often set ourselves apart from Nature, for much of our behaviour appears to run contrary to the laws by which all other animals lead their lives. But one cannot help wondering, as Man, tree shrews and all the primates between them move inexorably towards the twenty-first century, a mere drop in the ocean of geological time, what trump cards the perpetually changing theme of evolution may have in store for us all.

Before we consider the main groups of primates, we must take a closer look at the tree shrews of South-East Asia. They are important because arguments rage as to whether or not they are actually primates. Certainly, they do not conform to our twentieth-century image of monkeys and apes swinging through the trees or beating their hairy, barrel chests with rage. But then nor do any of the 'lower' primates whose position on the evolutionary ladder to Man is undisputed. One of the exciting facts about tracing evolution through time is that the further back we go the more difficult it often becomes to relate the 'beginnings' to the 'end product'. Man did not evolve from tree shrews. All we can say for certain is that tree shrews were among the first mammals to get off the ground and that Man eventually evolved from tree-dwelling primates which returned to the ground only relatively recently. The 'jump' from tree shrews to Man is indeed colossal and it could never have even started to happen without conditions which demanded that some of the tree shrews went higher into the trees and underwent certain physical adaptations for living at that higher level.

Gradually, the differences between the various groups of tree shrews became more marked and after many generations the predominantly tree-dwelling groups had little in common with their 'ancestors'. Thus we end up with a gradual progression that leaves in its wake different kinds of monkeys as it moves slowly through time.

Much of the evidence for the evolutionary changes which took place among the primates comes from fossils whose age it is possible to work out. Additional evidence is provided by those animals which did not change but still managed to survive in almost their original form to the present day. But if we can confidently trace primate evolution back to the earliest prosimians, or 'lower' monkeys, then we must also go one step further. The prosimians did not suddenly appear in the trees. Before them were many insect-eating, ground-dwelling mammals suffering from intense competition between themselves for food and for places in which to live safely. Little wonder, then, that some of them turned to the emerging forests and undergrowth which would have provided plenty of insects at a higher and safer level. Of course they were limited in what they could do. They had long claws for scratching around in soft earth and long sensitive noses, adapting them more for smelling than seeing. They must have clambered around the low vegetation, using their claws to get a grip where hands would have been so much better. But the fact of overwhelming importance is that they made it 'off the ground', not very high but very successfully in contrast to their ground-dwelling ancestors and immediate competitors. They paved the way for Man to evolve millions of years later and the incredible thing is that some of them are still alive today and living much as their ancestors must have done in those primeval forests some 70 million years ago.

Today, tree shrews are found only in the lower growth of the dark forests of South-East Asia, including India, Malaya, the Philippines, and many of the large South Asiatic islands. They are small, agile and so reminiscent of squirrels that the natives in many areas call them both by the same name, *tupai*.

All the tree shrews retain many of the features of primitive insectivorous mammals. Their limbs are of roughly equal length, indicating that they never evolved a predominantly two-legged gait either on the ground or low down in the trees where they seem to be equally at home. Their hands and feet still have claws whereas the claws of the later primates developed into shorter nails on the end of sensitive finger and toe pads, and they lack the fully opposable thumb and big toe although these digits are sufficiently independent to suggest a trend towards the full opposability of the advanced primates.

In their attitude towards their babies they are also unique. The mother leaves her babies when they are still very young and helpless, returning briefly at intervals of about two days to feed them. Parental devotion had obviously not yet

been fully developed at this early stage of primate evolution, perhaps because of the dangers of living so close to the ground. It may well have been safer to leave the young alone rather than advertise their presence to predators by remaining close to them. More advanced parental care can be observed in the later primates, whose young are born either higher up in the safety of the trees or down on the ground to parents who are physically capable of defending themselves.

The commonest tree shrews are members of the *Tupaia* genus and they range in body length from about 15 cm (6 in) to about 22 cm (9 in), not including their tail which accounts for roughly half their total length. They weigh between 100 gm (3½ oz) and 200 gm (7 oz) and are coloured dark on top — in various shades of dull red, green and brown — and pale to white on their bellies. They are daytime species and many spend only part of their time in the trees.

The Pen-tailed Tree Shrew belongs to a different genus in which there is only one species. It is unique among the tree shrews because it is the only truly nocturnal and tree-dwelling member of the family. It is about the same size as its relatives but is more thickly and softly furred on its body. Its tail, however, is completely naked except for a tuft of white hairs at the tip. This extra-ordinary adornment may well serve as some kind of signalling device between members of the same species on dark forest nights. The tail is such an outstanding feature of this small rat-like animal that its common name is derived from it. With its enlarged hands and feet, which bear sharp, pointed claws, the Pen-tailed Tree Shrew is an expert climber and nests in tree-holes as much as 18 m (60 ft) above the forest floor. The nest is made mainly of leaves, mixed with twig and wood fibre.

When they do venture down onto the ground — they have been trapped there but rarely observed — Pen-tailed Tree Shrews hop rather than walk and this may well be an important factor in finally deciding whether tree shrews are primates or not. Their closest descendants, especially the bushbabies, also hop on the ground even though they are far more spectacularly adapted to living in trees than are any of the tree shrews. Whatever the true position of the tree shrews, they need to be studied more thoroughly by scientists who, presented with what is perhaps a perfect bridge between two evolutionary pathways, may never fully agree on which side they belong.

ABOVE
There are many arguments for and against the inclusion of tree shrews among the primates. The most obvious criticism is that they do not look at all like the primates — though they certainly are not shrews or squirrels! — even though such striking differences are not always endorsed by studies of internal structure. Although they do not possess the stereoscopic vision provided by the evolution of forward-facing eyes, the actual structure of their eyes is of the sort that could easily have given rise to that of more advanced primates.

BUSHBABIES, LORISES AND TARSIERS

If aware of a predator, a Potto will normally 'freeze' until danger has passed by, but in the event of an imminent attack, it lowers its head between its front legs, presenting the back of its neck to its assailant. In a head-on confrontation on a thin branch the predator has no alternative but to attack from that direction and, seemingly encouraged by the attitude of the Potto, it moves in for the kill. But however vulnerable the Potto may seem, it is in fact very well protected by several bony projections on the back of its neck which are covered by a thick layer of skin. Thus its spinal cord is hidden quite deeply from penetrating teeth, should they get that far. In addition, there are several hairs on the Potto's neck which project beyond the normal length. These are highly sensitive and act as critical factors in the repulsion of an attack for, in adopting such a defensive posture, the Potto has buried its head between its legs and cannot see very well. If it relied on smell alone, it would certainly know where its attacker was but it would have only a vague idea as to how far away it was. The projecting hairs are inevitably touched as the predator moves stealthily forward to get a good grip on its prey and the Potto reacts violently by bringing up its head and shoulders. The hairs are short enough to guarantee that the predator is close enough to receive a nasty blow from the rearing Potto and remembering that the whole drama has taken place on a thin branch, it is usually sufficiently surprised and stunned to lose its grip and tumble to the ground.

PREVIOUS PAGES
Photographed on its lethargic journey through a night-enshrouded rainforest in Malaya, a Slow Loris displays many of the characteristics which befit its rather secretive life-style. Its large, round eyes are well adapted to seeing as much as possible in the dark and it lacks the tail found among other, more active, primates that jump and run quickly through the trees. The hind feet clearly demonstrate the gripping power of the Slow Loris's toes — the big toe being set well apart from the remaining toes to enable this 38 cm (15 in) primate to secure a firm hold on the branch. This physical advantage is further enhanced by an arrangement of muscles in both hands and feet which close round a branch in such a powerful way as to make it almost impossible for the loris to lose its grip.

Having put the tree shrews firmly on the bottom rung of the evolutionary ladder which ends in Man, we can now climb a little higher. The territory of the prosimians, which include all the 'lower' primates before the true monkeys, once spread out across North America, Europe, Africa and South-East Asia. Although they do not live in all of these areas today, there is much fossil evidence to suggest that they were then a very widespread and successful group. Of course, during this time of the Eocene and later the Oligocene, they were the most advanced primates alive and they experienced little competition for as much as 20 million years. With no immediate competitors and with endless forests to colonize, they were able to flourish and expand.

The tropical forests we find in the world today are vastly reduced in size from the great tracts of uninhabited vegetation of the Eocene days. Not only have further changes in world climate reduced them in size, thereby increasing competition between all the animals that want to live in them, but more recently and very significantly, Man himself has begun to cut them down to clear paths for cultivation as he too spreads throughout the world. At the present rate of clearance, one cannot help but wonder how many years from now will see all the big forests completely destroyed. If this happens, many animals that we have all grown to know and love will disappear for ever. Animals that have existed for millions of years just cannot evolve fast enough to adapt to such a major catastrophe. But, for the time being at least, we can still enjoy the company of several different kinds of prosimians in Africa and in South-East Asia.

When the first prosimians evolved, they must have done so gradually and it might be reasonable to assume that they resembled the tree shrews which exist today in South-East Asia. They would have been small, with long tails, and they would have had short limbs ending in five digits. Their eyes would have been set on the sides of their heads, their nostrils long and pointed, indicating that their sense of smell was still more highly developed than their eyesight. But they undoubtedly had many advantages over their ancestors and as time carried them out of the Palaeocene and into the Eocene era, they spread across the world, changing as they went.

The different kinds of prosimians that evolved did not necessarily represent better models than those that went before them. Nature, in a sense, experiments, often producing different models in different places at the same time. Some will evolve along lines that will eventually give rise to more advanced forms, while others will progress no further and yet still survive. Others will die out altogether, becoming extinct, and it is only by finding their fossils that we are able to appreciate how and when they may have lived, what they looked like and so on.

Although the fossils that have been found so far tell us that the prosimians were once widely distributed throughout much of the world at a time when vast tropical forests abounded, we must also remember that even the continents of the globe have been slowly moving. When the first prosimians evolved, the stretches of water that we now see as such formidable barriers to their dispersal may not necessarily have existed.

Because there are so many diverse prosimians living today, divided into distinct groups, we will deal with them in separate chapters. Here we shall look at the lorisoids which live in the tropical regions of Africa and South-East Asia, and the following chapter will cover the lemuroids which are found only on the large islands of Madagascar and the neighbouring Comoro Islands in the Indian Ocean.

We can divide the lorisoids into three distinct groups: the lorises themselves inhabit tropical Africa and southern Asia, the galagos, or bushbabies, are widely distributed in Africa south of the Sahara, and the rare tarsiers are confined to the forests of South-East Asia.

The lorises are renowned for their slow and deliberate movements — their name is derived from the Dutch word *loeres* which means 'slow-witted'. This is hardly a fair label to attach to these little creatures which are in fact capable of very quick movements, especially if they are disturbed or when they are in search of food. It is important for them to move slowly when they are hunting because they feed mainly on insects and small birds, and the last thing they want to do is frighten their prey away before they reach it. So they approach very carefully and then make a final, rapid lunge at their chosen meal.

All the lorises are nocturnal, a way of life that was forced upon them largely by the emergence of many day-time predators and the even more advanced primates

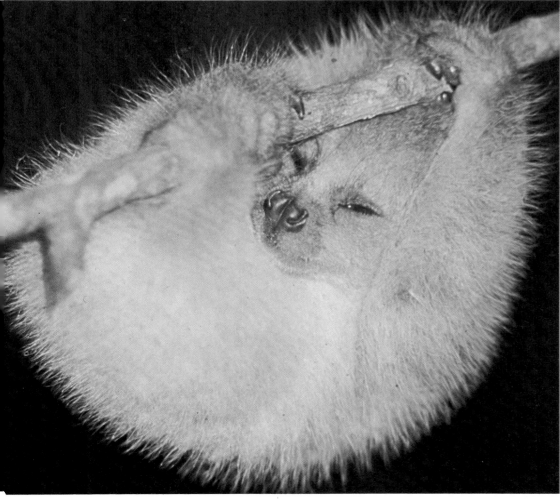

LEFT
The young Potto clinging defencelessly to the branch of a rain-forest tree is the sole infant born to its mother during the year. In the wild, births have been recorded only between the months of August and January and although occasionally twins may be born, this is a phenomenon which has been observed only among captive animals. The baby Potto is dependent upon its mother's milk for about the first two months of its life, but it will be weaned by March of the year following its birth. By that time the forest trees will be laden with fruit and the young Potto's needs will be fully catered for.

When faced by a predator the Angwantibo defends itself in a rather different way to the Potto. It spends quite a lot of its time on the ground and is therefore susceptible to more terrestrial carnivores. When it senses danger it will readily 'freeze' on the spot, but it often climbs into the closest tree and then waits, moving higher if pursued. If it is finally attacked it also draws its head between its front legs, but its main area of defence is its tail. This has on it a smooth black patch surrounded by paler-coloured hairs and when these are erected, a circle is formed. As the predator approaches — unlike the Potto, the Angwantibo maintains a constant watch on its attacker — it moves its tail and the predator's attention is diverted by the black and white markings. At this point the Angwantibo shoots out its head from under one of its front legs and administers a bite to the piece of body closest to it. In the ensuing recoil of pain and surprise, either or both of the animals lose their balance on the branch and one more Angwantibo mortality has been successfully averted.

— notably the monkeys — which were beginning to dominate all the 'lower' primates in the trees for both space and food.

The lorises are unique among primates in one particular part of their anatomy. They have developed more vertebrae in their backs, giving them considerable advantages when they are clambering around in the trees for they can twist round above and below branches and stretch out on their long limbs to get to places which might look way beyond their reach. Their tails are either absent or very small and the second digit on their hands is greatly reduced in size. As with all the prosimians, lorises scent-mark trees with their urine, which they normally wipe from their hands onto branches at strategic points in their nightly journeys. From these markings individuals can identify each other and, if necessary, take appropriate evasive action without coming into direct conflict with each other.

The weirdly named Potto and Angwantibo are lorisoids which are confined to the equatorial forests of central and western Africa, where they pursue an almost totally nocturnal existence among thickly leafed trees and vegetation. The Potto is the larger of the two animals, reaching a size of about 33 cm (13.2 in) as opposed to about 25 cm (10 in) for the Angwantibo. Because of this, there is a critical difference in much of their behaviour. The heavier Potto is unable to gain access to food reached by the light branches available to the lighter Angwantibo; this also applies in reverse for the bulkier Potto can grip branches too large for the Angwantibo, which easily loses its grip and falls to the ground.

Both Pottos and Angwantibos feed occasionally on small mammals and birds but their major source of food is fruit and insects which they locate by smell in the thick, night-enshrouded undergrowth. On nights bathed in the increased light of a full moon, they probably use their eyesight as well but this is not as critical a factor in their existence as it was for their day-living ancestors. They clamber slowly but deliberately through the branches and vines, ever alert to the slight rustlings that might betray the approach of a hungry predator.

Because of their slow movements, neither of them is able to make off at high speed, and to protect themselves they have evolved ingenious means of coping with predators. Many fierce, carnivorous mammals roam the forests at night, ever ready to snap up a quick meal. Leopards, civet cats and genets and other members of the mongoose family are all their natural enemies and yet, when they come up against a Potto or an Angwantibo they only rarely succeed in making off with their prey. Pottos and Angwantibos are among the few prosimians which 'stay put' and defend themselves. Their cousins, the bushbabies, for example, are of a more nervous disposition and at the first sign of a predator will leap off through the trees to safety.

There are no more than three species of loris in existence today outside Africa. They are the Slender, from the Indian sub-continent, the Slow and the Lesser Slow Loris, which live in South-East Asia. They tend towards being slimmer than their African relatives and probably add more animal flesh to their staple insect diet.

The Slender Loris ranges through much of the forested regions of southern India and Sri Lanka (formerly Ceylon) from sea level to about 1800 m (6000 ft). It is entirely nocturnal and spends the day on a shaded branch, curled up in a fluffy ball in the middle of which hands and feet, with their opposable thumbs and big toes, grip tightly to secure a firm hold.

The Slow Loris enjoys a far wider and more northerly distribution than its Slender relative, being found in Assam, Burma, Thailand, Indo-China (where the Lesser Slow Loris also lives), certain Malayan states and the East Indian islands of Sumatra, Java, Borneo and the Philippines. Throughout this quite extensive range, it is not very common and its furtive, nocturnal habits make it difficult to observe.

Throughout their ramblings for food, all the lorises are solitary. This does not mean that they are not sociable, but rather that they do not move around at night in groups. They undoubtedly meet up with each other at various points along their routes and react to each other according to their status within the group to which they belong. Such status, forming a loose hierarchy, depends upon the age and sex of each individual. This is especially true of the bushbabies, which are highly active creatures and cover quite a lot of their territories (perhaps more appropriately called their 'home range' because they are not, strictly speaking, territorial) on any one night. In contrast, the chances of Pottos and Angwantibos meeting up during the night are considerably reduced on account of their lethargic progress through the trees.

The bushbabies, which form a separate group of the loris family, are widely distributed in Africa south of the Sahara desert. There are six different species, occupying the rain forests of the equatorial basin and the more dry and open woodlands to the north, south and east. They range in size (including their tail which accounts for slightly more than half their total length) from 77 cm (31 in) for the Thick-tailed Bushbaby down to the minute Dwarf Bushbaby, only about 28 cm (11 in) in length.

The expert at living in dry areas is the 39-cm (15-in) Lesser Bushbaby. Along with its Thick-tailed relative, this species has successfully broken away from the equatorial basin, but to a far greater extent. Today its range extends throughout much of the savannah woodlands south of the Sahara and is excluded from only the southern tip of the huge continent.

Identification at night is facilitated by trapping and then harmlessly marking each animal on its tail. The observer simply looks up the combination of tail marks (which grow out after a while and have to be renewed) and is provided immediately with the pre-recorded age and sex of the bushbaby he or she is watching. Armed with such information, the scientist can build up detailed records of social organization within the bushbaby population. By following one individual throughout the night, information on its nocturnal ramblings and feeding habits can be recorded and its relationships with all the other bushbabies it meets more clearly understood.

To study the breeding habits of nocturnal animals obviously poses quite a problem for scientists. Much detailed information can be gleaned from studying animals in captivity, but in the forests much of this behaviour is missed and scientists must find either pregnant females or those with young in attendance before they can attempt to work out breeding cycles. Generally, the African lorisoids breed at a time when food levels are rising to a maximum. Among those

ABOVE
This tree-hugging tarsier, seen in an uncommon daytime photograph, shows many of the features that make it so unique among the prosimians. Its eyes, superbly evolved for a nocturnal existence, are the largest, in relation to their owner's size, of all primates and its hindlegs are by far the longest. Its elongated tarsal bones (those of the ankle) are so specialized for jumping and landing, as it moves from branch to branch in frog-like fashion, that it is from them that the name 'tarsier' is derived. Its tail, long and smooth with no more than a tuft at its tip, gives added body support against the tree trunk and is probably used as a kind of mid-air rudder to assist in any required change of direction. But perhaps the tarsier's most fascinating feature is its ability to rotate its head through 180 degrees. No other primate can manage more than 90 degrees and yet the tarsier's speciality, which it shares with the owls, is one of the reasons why this harmless little reminder of a bygone era is viewed with suspicion by many of the superstitious inhabitants of the islands on which it lives.

As with many of the lorisoids, Demidoff's Bushbaby builds a nest of fresh leaves in which to care for its young. Although they fall apart as the leaves die, the nests may also be used by adult bushbabies in the non-breeding season as a sleeping place during the daytime. At such times these normally elusive creatures can be photographed at quite close quarters, although the one shown here, photographed in the Samunya Forest in Uganda, is reacting quite viciously to the intruding camera.

The young Slow Loris pictured here is not yet sufficiently aware of the dangers of being surprised in the jungle to respond in the normal fashion of quickly curling up into a tight ball of protective fur. It will have been the only baby born to its mother after a pregnancy lasting some 90 days, and it will probably not have left her very long ago. When the young are born they attach themselves to their mother's fur — back or front, above or below, depending on how she is travelling at the time — remaining as passengers until they are almost as large as she is.

that live in the tropical forests, in which there is a fairly continuous supply of fruit, breeding is likely to be spaced throughout the year with no more than seasonal peaks, but for the Lesser Bushbaby this is not the case. In the arid savannahs of South Africa where it lives on insects during the wet summer months there are two breeding seasons, one at the onset of the rains in November and a second towards the end of the rains in February. During the dry winter there are virtually no insects for the bushbaby to eat, but it appears to have overcome such an outstanding problem by relying solely on the gum which flows out of some species of tree in the acacia-dominated woodlands.

The irony here is that the gum flows only during the wet, warm summer months when the bushbabies are feeding on an ample supply of insects. During many of the winter nights temperatures fall dramatically, the insects disappear and the gum freezes solid. Only by scraping with their back teeth and by licking with their tongues are the bushbabies able to stay alive through several months of the year. The gum is evidently highly nutritious for it is probably their sole source of food during this dry period, but still most individuals lose up to a fifth of their body weight before they can gorge themselves on summer insects once more.

The tarsiers of South-East Asia are a particularly interesting group, consisting of only three recognized species from the island regions of Sumatra, Borneo, Celebes and the Philippines. In their forested isolation, they appear to have remained unaltered for as much as 50 million years of primate evolution and their ancestors were quite likely the very prosimians whose fossil remains have so far been found in European and in North American Eocene deposits.

Tarsiers feed largely on insects, but they will also seize eagerly upon nestling birds and small rodents, many of which may well be their equal in size for, tail apart, they are themselves no larger than a rat. They breed throughout the year, often carrying their single young by the scruff of the neck in the same manner as the bushbabies. So transportation leaves all four limbs free and progress from tree to tree is not impeded.

PREVIOUS PAGES
Held firmly on a branch by its powerful feet, a Slow Loris has its hands free to hold on to a cricket on which it is feeding greedily. Perhaps the result of a patient, slow-motion stalking session, the prey is devoured even at risk of allowing a photographer to approach closely. So strong is the grip of the hindfeet that the loris quite frequently hangs upside down from a branch and gathers food with its hands. The Slow Loris feeds on insects most of the time but it will readily eat fruit and other parts of plants. Occasionally it will stalk and successfully capture small mammals and birds as they sleep in the trees at night. The short, thick fur of these lorises varies a lot on their back — some are pale in colour while others are dark brown — but they are almost all equally pale below.

ABOVE
Bushbaby males compete among themselves for females and the outcome is normally determined by age and therefore mating priority. The female is pregnant for about three months, during which time she builds a nest of leaves in which to give birth to one, two or even three babies. She leaves her young undefended, returning at frequent intervals through the night to suckle them. Her visits decrease as the young grow. Every so often she picks up her young by the scruff of the neck and removes them to another nesting site not very far away. This is an ideal way of preventing the accumulation of obvious tell-tale signs that helpless young are present, and it serves admirably to keep predators from being presented with a free meal in a rather exposed habitat.

RIGHT
The Thick-tailed Bushbaby is confined largely to equatorial Africa, although it has to some extent broken away from this typical lorisoid habitat and has extended its range even farther. It is today also found in the isolated forests which hug the steep contours of river gullies in mountain ranges. Wherever it lives, it feeds on a mixed diet of fruit (when in season), insects and the gum from certain trees. Typical of all the bushbabies, the Thick-tailed species can move its ears independently of each other in order to locate more precisely the whereabouts of an insect moving in the darkness of night. It is from the plaintive, child-like calls of this species that the bushbabies derive their English family name.

The Lesser Bushbaby is one of the most endearing creatures alive. Its small size and large appealing eyes, combined with a general fluffiness, make it a common pet in Africa where some individuals have been known to live for more than ten years in captivity. In the wild, because of the hardships they are continually faced with, their natural lifespan is probably no more than three to four years. This bushbaby has been quite well studied in South Africa where a two-year study is currently being conducted on a known population of some 25 individuals. The animal shown here is a member of the study area in the Northern Transvaal.

BELOW
The Lesser Bushbaby is about the most agile of all the prosimians and it can make its way through the tops of even the most thorn-laden trees at high speed. Having forsaken its dense forest habitat for the more open woodlands, it has developed enormously large hind leg muscles to overcome its increased vulnerability to predators.

LEMURS
THE PRIMATES OF MADAGASCAR

Two female Black Lemurs balance precariously on a slender branch while one of them eats a banana. This species exhibits a phenomenon rare among mammals — that of colour sexual dimorphism. While these females are reddish brown with white bellies, the males are completely black. This is also an interesting species because it eats mostly fruit and leaves in the wild, and yet captive individuals will readily eat birds, cracking open their skulls and consuming the brain as though it were no more than a hard nut from a tree! Black Lemurs live in forested regions and travel in groups in which the females form the advance party and the males bring up the rear. Youngsters are summoned to their mother's side by rapid grunting calls which, appropriately, also serve as alarm calls. Should danger persist, these calls become louder and more frantic until they are almost continuous. Black Lemurs are diurnal and are most active during the early morning and late afternoon when the sun is quite cool. They resemble the Ring-tailed Lemur in many aspects of their behaviour, although unlike them they are expert tree-dwellers and can leap as much as 8 m (26 ft) from branch to branch.

ABOVE
Crouched on a fallen log and looking more bear-like than monkey-like, a tailless Indri does not give the impression of being well adapted to life in the trees. In fact, the powerful hindquarters are highly flexible springs with which it is capable of launching itself many feet from tree to tree. The Indrises show a great deal of individual colour variation — generally they are patterns of black, grey and brown but not infrequently they may be almost entirely black or sometimes white. Whatever the colour, the rump is nearly always white bordered by black. This could be useful in display and may be part of courtship behaviour.

For about 35 million years, the island of Madagascar (now the Malagasy Republic) has been separated from the main African continent. Since then, the African mainland has seen many changes in its animals. We have already seen which of the prosimians are still to be found there and soon we will come across a myriad of monkeys and apes that evolved later on.

It is, however, worth noting here once more that as these monkeys and apes evolved, they invariably took over from the prosimians, forcing many of them into extinction or into adopting new and non-conflicting life-styles in order to survive. It is a credit to the prosimians that exist today that they were in fact able to cope with this onslaught. It is perhaps more interesting to note that not a single lemuroid prosimian survived for any length of time outside this isolated Madagascan region in the Indian Ocean.

If we turn again to the fossil record, we see that the lemuroids once flourished in Africa itself and even as far afield as North America. We might reasonably assume, therefore, that if Madagascar had not been severed from the mainland there would be no living lemuroids in existence today. There is a school of thought that suggests — using fossil records to date as evidence — that the lemuroids were not actually isolated on Madagascar at the time when Continental Drift caused it to be separated from the African mainland. If this is true then zoologists must find another explanation for their presence on the island today; and the only possibility is that some of them drifted across from Africa on floating vegetation and subsequently evolved into the different species that we find today. So, on this island we have a unique situation — almost a living museum — which has greatly enhanced our understanding of primate evolution.

Lemuroids are often referred to as 'living fossils' for had this particular geological event not happened, they would almost certainly have died out many millions of years ago. As it is they survived, oblivious to the rest of the world and under such privileged conditions that they spread throughout the island's lush forests, evolving into a number of different forms as they went. The reasons for this isolated survival are fairly clear. Protected from the mainstream of primate evolution, the lemuroids spread rapidly and safely. They met and adapted to different habitats and in them they underwent specific changes. In this respect they were fortunate for when they were separated from the African mainland they took very few predators with them. Later on they were spared the intimidating evolution of monkeys and apes. Without these critical 'eliminators' and 'competitors', some lemuroids became nocturnal — or perhaps *remained* nocturnal if they showed that inclination before they were isolated — while others remained or became diurnal, for this was to remain an unchallenged way of life for them to explore quite safely. Virtually all of them live in the trees, the only exception being the Ring-tailed Lemur which spends a lot of time on the ground where it walks on all fours.

The lemuroids range in size from the diminutive Mouse Lemur which is no more than 12-13 cm (5 in) long to the largest prosimian of all, the Indri which measures about 1.2 m (4 ft) from head to toe. The Indri also happens to be exceptional on Madagascar because, unlike all of its relatives there, it does not have a tail. This factor immediately classes it as ape-oriented and, like them, the Indri is diurnal and lives in small social groups. But these comparisons with the 'higher' primates should not be carried too far. According to evolutionary logic, the Indri may have returned to the ground sometime within the past 10 million years, lost its tail because it would have been a liability not a help at that level and then returned to the trees where today it is an expert leaper and seems quite at home. David Attenborough, in his book *Zoo Quest to Madagascar,* relates how amazed he was to see how well the Indri coped high up in the trees without a tail. In fact, he was so convinced by their aerobatics that he could not imagine how they had returned to the ground, lost their tails and then gone back up the trees without undergoing other radical changes at the same time. Islands often produce bizarre situations such as this one.

Today, the Madagascan region supports three distinct families of lemuroids. There are the true lemurs themselves containing about sixteen species living in predominantly woodland areas. Apart from the exclusively diurnal Ring-tailed Lemur, they are largely nocturnal in their habits, spending the daylight hours tucked away in low undergrowth, trees or bamboo thickets. Family groups consist of up to twenty individuals which are often so active that it is difficult to count

their numbers accurately.

Some of the lemurs store fat in their bodies, especially their tails, and then spend the hot dry season sleeping and living off these stored reserves which are absorbed into their bodies at a slow rate. Sleeping through the summer months this way is called estivation and is the opposite of the commonly known habit of sleeping during the winter (hibernation).

Most lemurs breed once a year — although some may do so twice — and they produce one to three youngsters after a pregnancy of around five months. The young are born open-eyed and are cared for by both male and female during the first few months of their lives.

Lemurs feed almost exclusively on leaves and fruit, foraging in the trees and occasionally on the ground, but they supplement this diet with insects whenever they can catch them.

The second family group contains the more weirdly named Indri, the Avahi and the Sifakas all of which have shorter noses than the lemurs and thus take on a more dog-like appearance. They seem to be more cumbersome in the trees than the lemurs but they are nevertheless ideally adapted, being expert climbers and leapers. Only the Avahi is nocturnal, and all of them may be found singly, in pairs or in small groups as they progress hand-over-hand through the branches of the trees. Occasionally they will rest with their arms and legs held open to expose

ABOVE
The magnificent Ruffed Lemur holds two records among the true lemurs. First, it is the largest of them all — a full grown adult measuring some 120 cm (4 ft) from its head to the tip of its tail, and, second, it is the only true lemur to build a nest for its offspring. Shortly before she gives birth, the female constructs her leafy nest and then pulls hairs from her body, especially from her haunches, to provide a warm and secure lining for her helpless youngster. The single offspring, although occasionally twins may be born, is weak and cannot climb at all well until it is at least five or six weeks old. A sudden increase in strength means that by eight weeks it can climb very well and manage to feed itself on vegetation.
Ruffed Lemurs travel in small family-type groups of two to four and occasionally emit loud roars followed by clucking sounds which fade away into the dead of night.

RIGHT
A typical view of Ring-tailed Lemurs in a
woodland clearing shows how distinctive
the black and white tails are from a
distance. This is an exclusively diurnal
animal and the prominent 'flags' must
serve as visual contact signs as the lemurs
make their way through tall grass or forage
among the woods. By such tell-tale signs
and occasional high-pitched calls the unity
of the group is maintained. Further contact
is established by both males and females
scent-marking trees, although this is also
part of the ritual of marking group
boundaries as well. Young lemurs are an
integral part of such a group as this and
they stay close to their mothers for at least
the first six months of their lives.

OPPOSITE
A close up of a male Ring-tailed Lemur
shows it waving its highly conspicuous tail
in the air. This gesture is actually part of an
intricate process by which males establish
their dominance over each other within a
group. The males possess scent glands on
their chests and on their arms and as they
secrete liquid from these they rub it into
their tails. Suitably drenched with the
pungent odour, the tail is raised and waved,
the smell being carried through the air
towards the rival male. These aptly named
'stink fights' are highly effective for
dominance is probably established by this
act alone, although fighting occurs during
the breeding season. The stronger animal
advances towards his rival who,
acknowledging his inferior status, backs
away and the score is settled without any
physical contact being made.

their underparts to the warming sun — an attitude which has caused them to be
labelled by natives as sun-worshippers! They feed exclusively on vegetation
which commonly includes leaves, pieces of bark and flowers and anything else
that they find botanically appetizing, either on the ground or in the trees.

As with the lemurs, they normally breed once a year, producing a single
offspring which is carried through the trees for the first few months of its life.

The third family contains but a single species, the bizarrely named Aye-Aye
which weighs about 2 kg (4½ lb) and, including its slightly longer-than-body tail,
measures on average some 100 cm (40 in). The Aye-Aye is exclusively nocturnal
and arboreal, spending the daylight hours asleep in holes in trees or in a nest
among the branches. At night they move around either singly or in pairs. They
are unique in having an extremely slender third finger on each hand which they
use extensively for feeding and for grooming. Of their breeding behaviour very
little is known, although it is likely that they produce a single young each year in
keeping with the lemuroid pattern. A leaf nest is probably built up in the trees for
the raising of the youngster.

A different kind of Aye-Aye lived on Madagascar about two or three million
years ago. Its larger size must have made it an awe-inspiring creature and it was
probably regarded with the same suspicion as its present-day relative which is
often considered a reincarnation of long-dead natives.

Clinging tenaciously to a branch in the
night, a Lesser Mouse Lemur seems to be
positively encumbered by its grotesquely
swollen and hairless tail. Such an
ugly-looking appendage may perhaps
hinder this lemur during the wet season
but during the hot, dry season it acts as the
sole means by which it stays alive. It acts,
quite literally, as a swollen larder of highly
nutritious fat, stored away when food was
plentiful during the wet season. The Lesser
Mouse Lemur, along with only one other
kind of lemur — the Fat-tailed Dwarf
Lemur — may go into hiding during the hot
and dry months of the year and spend most
of the time asleep. The amount of food its
body needs to survive is greatly reduced by
this state of inactivity, and the fat from the
tail is gradually absorbed to cater for its
minimum metabolic requirements until
the following wet season. The Lesser Mouse
Lemur is possibly the smallest of all the
primates (excluding the rather under-rated
tree shrews). Adults may weigh as little as
40 gm (1½ oz) and, including their lengthy
tails, measure upwards from 30 cm (12 in).

The Fat-tailed Dwarf Lemur employs the
same habit as the Lesser Mouse Lemur of
storing fat reserves in its greatly distended
tail to help it through the hot, dry summer
months when it hides away and becomes
rather torpid. During the wet season, it
feeds almost entirely on fruit which is not
available all the year round. Two young are
normally born just before the wet season
sets in, so that by the time they are weaned
and can move away from their twig and leaf
nest in search of their own food, the
surrounding fruit is ripe and readily
available to them. Occasionally the mother
will move them to another hiding place by
carrying them in her mouth.

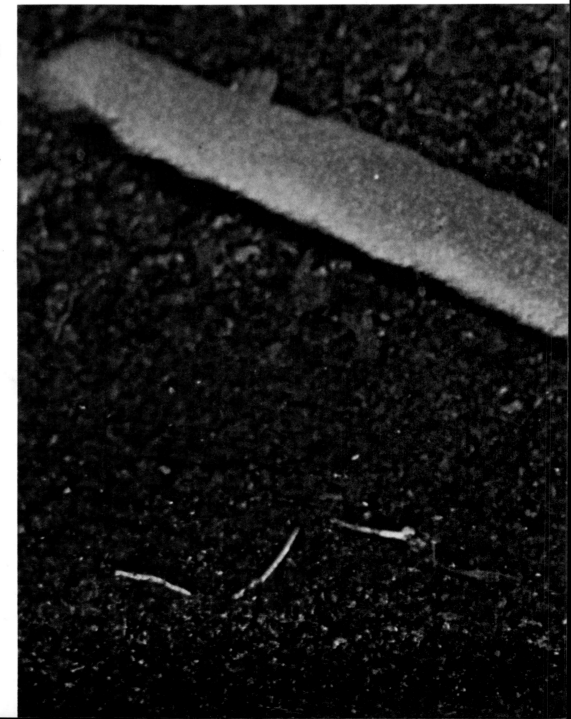

The aptly named Grey Gentle Lemur is nocturnal and spends the daylight hours sleeping in low undergrowth and bamboo thickets. Bamboo forms one of its major sources of vegetation in its diet and its teeth, except its grinding molars, are equipped with sharp cutting edges to help it tear through the tough outer covering to reach the soft pith inside. Gentle Lemurs usually move around singly although two or three together would not be uncommon especially during the months when young are dependent on adults. Both sexes have a curious scent gland on the inside of their wrists. In the male this is rough, bare skin covered with spines while in the female these spines are replaced by softer hair.

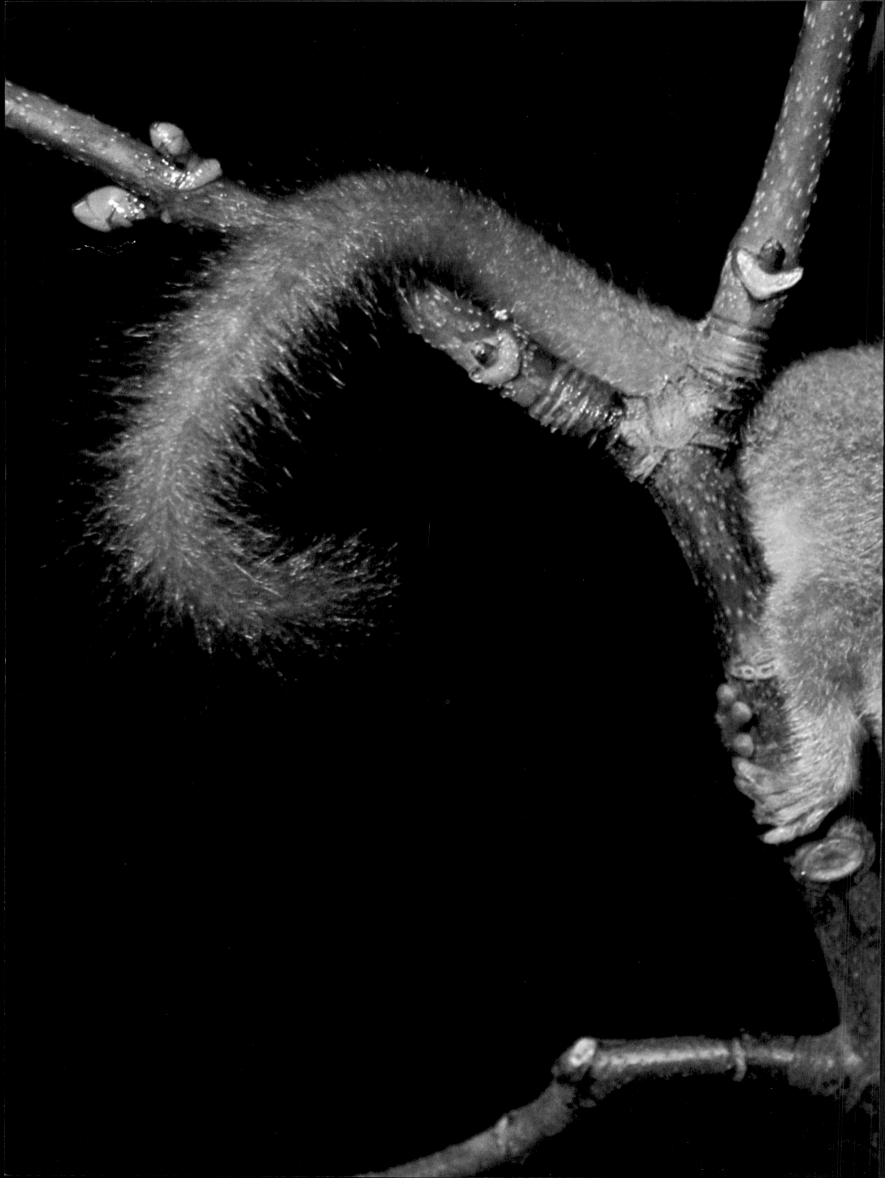

PREVIOUS PAGES
This Brown Mouse Lemur is a close relative of the Lesser Mouse Lemur but lives in rain forests. To some extent it stores fat in its hindquarters and its tail, but the habit of estivation is not a regular feature of its life. Mouse Lemurs are nocturnal and shelter in tree-holes during the day. Occasionally, abandoned nursery nests of leaves and twigs placed high up in the fork of a tree may provide shelter during the day. As with most nocturnal animals which are partially blinded by the glare of the sun, Mouse Lemurs have a reputation of being sluggish during the day. But as the sun sinks below the horizon, they become highly active, moving swiftly through the trees on all fours, leaping expertly from branch to branch and using their stout tails to maintain their balance. They have a reputation for being shy and aggressive, uttering shrill whistles of alarm when disturbed or when they fight among themselves.

RIGHT
With her few month old offspring clamped firmly on her back, a female Verraux's Sifaka makes a hasty retreat up a tree in Madagascar's savannah woodlands. When they do venture to the ground, Sifakas hop on their hindfeet, waving their arms in the air to maintain their balance. This may explain why they tend to pick up food in their mouths rather than their hands, although they feed mostly in the trees where there is a ready supply of fruit and leaves. In the early morning before the heat of the sun has taken effect, Sifakas satisfy their thirst by licking fresh dew from the leaves.

LEFT
An exclusively vegetarian Avahi clings vertically to a tree trunk and shows off its enormous, thick tail. This individual belongs to the northwestern population of Avahis in Madagascar. Although there is one species only, another form is known from the forested areas on the eastern side of the island. Avahis are solitary, nocturnal lemuroids which makes them difficult to observe. Like the Sifaka, they show considerable colour variation and they also hop on the ground waving their balancing arms in the air. As with the other members of the indrid family, the single young are born towards the end of the dry season so that by the time they are ready to move around on their own among the branches, the wet season has begun and plenty of leaves and buds are available for them to feed on.

LEFT
With a prodigious leap high above ground, a fully grown Verraux's Sifaka sails across from one Didiera stem to another nearby. As it leaps across with power generated solely by its hind quarters, the Sifaka holds its arms and legs forward so it grasps equally with all four limbs as it lands. Until scientists realized that these leaps of twenty feet or more really were the result of muscle power alone, they were convinced that the arm and leg spreading habit stretched out the loose skin between its limbs, enabling it to glide such distances between trees.

37

Skulking close to the ground in a dense bamboo thicket, an adult Aye-Aye displays a shaggy coat and bushy tail as it searches for a nocturnal meal. The diet consists of insects, grubs, bird's eggs and fruit. The elongated third finger — not visible in this picture — is used especially in catching wood-dwelling larvae. The Aye-Aye will tap branches until it has located the exact position of its prey and then insert this strange finger into the wood to catch the larva. If the hole is too small, it uses its sharp teeth to enlarge it. This third finger is also used as a special grooming comb and all the fingers combine to assist the Aye-Aye in drinking. It will submerge its hand in water and then draw its fingers one at a time through its mouth.

BELOW

There is just a single species in the genus to which the Sportive or Weasel Lemur belongs. It is widely distributed throughout Madagascar's tropical forests and is also found on the small island of Nosy Bé off the northwest coast. The Madagascan population nearly always spends the day asleep in tree-holes, but it is interesting to note that the small island inhabitants generally curl up in a ball on an exposed branch. Quite likely, this habit results from two factors. First, there are no natural predators on the small island and, second, temperatures are slightly higher than on the nearby larger island which is further away from the equator. These little lemurs are solitary, very shy, arboreal and nocturnal, all of which combine to make them a rare sight even though they are widely distributed. Consequently, not a great deal is known about their behaviour. Their breeding habits are well documented and the single young is born after a pregnancy of about 135 days. The first few days of its life are spent concealed in the warmth of its mother's abdominal fur and although it is not fully weaned until it is about six months old, it begins to feed on its mother's food — fruit, leaves and probably a few insects — when it is about five or six weeks old. It will follow its mother around for as much as a year but when she becomes pregnant once more she loses interest in her self-sufficient youngster of the previous season.

MONKEYS FROM THE NEW WORLD

The first group of monkeys comes from the New World — from Central and South America. In this region monkeys are found only from Mexico to Argentina and, again, such a distribution is limited almost entirely to the extent of tropical rainforests. Unlike their Old World counterparts, these monkeys are confined almost exclusively to a life in the treetops. With long raking limbs and the evolution, among some of them, of the unique prehensile tail which acts as a fifth grasping limb, they are superbly adapted to a life high above the forest floor. At such heights they feed almost entirely on leaves and fruit.

New World monkeys are so different from the rest of the monkeys that the only explanation of their existence seems to be that they evolved quite separately from them but probably at about the same time — some 45 to 35 million years ago during the late Eocene and early Oligocene periods. The fact that all the monkeys share many anatomical and behavioural characteristics indicates that they evolved from a common prosimian ancestry, but their differences, and the additional fact that there are no apes in the New World today, suggest that they did so quite separately. This sort of situation, not uncommon throughout nature, is called parallel evolution and in this particular example the successful division between the New and the Old World monkeys was probably completed about 30 million years ago.

Fossil records of New World monkeys are scarce, but those that have been found indicate that they once lived in the Caribbean on the island of Jamaica and also down on the southernmost tip of South America, where the remains of a close relative of the Howler Monkey have been found. The fluctuating distribution of primates has always been closely linked to the expansion and contraction of the world's major forests, from their worldwide boom 60 million years ago down to their present-day distribution in equatorial regions.

Bound to this life in the trees, there has been no need for the New World monkeys to develop the marked size differences as shown among the terrestrial Old World Monkeys. Males are also far less aggressive, for obviously an arboreal existence poses far fewer problems than does one on the ground. There is an ample supply of food, water can be drunk in the form of fresh dew on leaves, and there is a noticeable lack of predators, although the larger pythons and some of the birds of prey constitute something of a daily hazard in their lives. They do not have cheek pouches and some of them have reduced thumbs to leave elongated fingers which act as hooks as they swing and leap through the trees. The main physiological differences between Old and New World monkeys are that the latter have no hard sitting pads, widely spaced nostrils, and no opposable thumb, although their big toe shows some similarities. But their prized possession is the prehensile tail. They do not all have this specialized use of their tail but those that have, especially the Spider and Howler Monkeys, use it to magnificent effect. It will easily support their full weight and they can swing from a branch by it alone, leaving hands and feet free to gather food. New World monkeys are quite small, generally weighing less than 10 kg (22 lb) as opposed to the 50 kg (110 lb) of the Mandrill, an Old World terrestrial specialist.

New World monkeys are divided into two groups, the cebids and the marmosets. The cebids usually live in small family units, producing a single youngster after a pregnancy of between 140 and 180 days. The cebids are all diurnal except the unique Douroucouli or Night Monkey. It is occasionally called the Owl Monkey because of its large round eyes. One male and one female usually travel around together with a single youngster, which is cared for predominantly by the male who releases it for feeding by his mate. The Douroucouli is one of the few monkeys which is monogamous and the couple's devotion to each other is often shown in the delightful tail-twining habit that occurs when they sit side by side on a branch.

The Titi Monkeys are similar to the Douroucouli in that they have a primitive brain but they are diurnal and live in small family groups which occupy such small areas that their densities in some areas may reach the figure of 500 per square kilometre (c. ⅓ sq. mile). Consequently their territories are hotly defended and when rival groups meet — and one imagines that this must happen quite frequently — there is much calling, hair-raising and tail-lashing to establish rights.

Among the Sakis, the males and females show some noticeable differences, for the males have a more strikingly coloured head than the females. They are

The Monk Saki is an animal of gentle disposition which will make a very affectionate pet if given the right conditions. It is a naturally shy and retiring primate and consequently has not been at all well studied in the wild. Its dark coat flecked with white makes it difficult to spot among dense foliage, despite its rather bulky appearance, although when it leaps from tree to tree its trailing bushy tail identifies it immediately. Its sharp, pointed teeth are expert at stripping berries from small twigs and for cracking nuts, as well as delicately skinning small mammals, birds and bats of their protective skins. Sakis will actively seek out such creatures, especially the nocturnal bats, by entering holes in trees during the day in the hope of catching them asleep.

Abandoned temporarily by its mother, a young spider monkey sits huddled on the bent fronds of a South American jungle palm. The bright white eye-rings visible in this youngster probably serve as a means of contact between individuals in the dark jungles where spider monkeys spend most of their lives high above ground among thin branches. But in such places these highly versatile monkeys have to be careful because they may weigh as much as 7 kg (15½ lb). In body length they may reach 62 cm (25 in) while their out-sized tail may increase this by as much as a further 90 cm (36 in). Spider monkeys usually run along the tops of branches with their heavy tails arched high over their backs. They travel in groups but remain in one area when they find an abundant supply of food.

RIGHT
The diminutive Squirrel Monkey is one of the great gymnasts of the New World rainforests. It will suspend itself from the thinnest-looking branch and twist its body round in search of succulent fruits, insects and even bird's nests, for it eagerly robs them of their eggs or nestlings. Squirrel Monkeys move through the trees — only rarely descending to the ground — in groups ranging from ten to one hundred individuals. They have a reputation for keeping themselves meticulously well groomed and clean, and yet they also have the habit of smearing themselves with a pungent secretion from certain glands, supposedly to keep potential predators at a respectable distance. No doubt such behaviour also leaves a distinct trail through the trees and serves to keep large travelling parties together as well as leaving clearly-defined trails to be followed later on.

diurnal and live in northern South American forests up to 200 m (c. 660 ft) above sea level. Their long, coarse fur protects them in the event of a sudden downpour of rain, a common feature of equatorial regions. Sakis eat fruit and other vegetable matter, as well as honey, small mammals, birds and bats from time to time. Red-backed Sakis live in dense forest on either side of the Amazon River, which has acted as such an effective barrier between the two groups that they have become sufficiently distinct to form separate species. This evolutionary phenomenon is known as allopatric speciation.

Perhaps the monkeys which are found most universally repulsive are the Uakaris. They have thinly distributed hair, virtually bald heads and pink mask-like faces with sunken and gloomy eyes.

There are three species of Uakari, the Bald, the Red and the Black-headed. They differ in fur colour but they all have the red face which turns even redder when they are excited or angry. Uakaris live in the tops of tall trees in swamp land and they rarely come down to the ground. Their tail is much shorter than that of other cebid monkeys. They live in small groups and their quietness enhances their rather sombre appearance. But for all their shaggy-furred 'ugliness', they are among the most expert of tree-dwelling monkeys, indulging in impressive acrobatics and leaps from tree to tree.

One of the most famous of the New World monkeys is the Howler Monkey. There are five different kinds which vary in colour from yellowish-brown to reddish-brown and black. They all have a remarkable vocal apparatus in which the hyoid bone, attached to the pharynx in the throat, is greatly enlarged. Thus, relatively small monkeys produce sounds which can be heard over very long distances. The 'howling' begins shortly after dawn and reverberates through the awakening trees to act as a signalling device between rival groups. They are not really territorial and this vocalization makes sure that they keep well away from each other as they feed during the day. Troops average about three adult males, seven adult females and the same number of developing youngsters. The males are slightly larger than the females and protect them in times of danger by calling vociferously and by breaking off branches and throwing them — or perhaps only dropping them — towards potential predators. The Howler Monkey has a truly prehensile tail which has a hair-free under surface which is so highly developed that it has prints like those of a finger at its tip. The grip of this appendage is so strong and secure that if the Howler Monkey jumped from a branch without releasing its hold it would not, as has actually been recorded, get very far!

Rather different from the Howler Monkeys are the two species of squirrel monkey. The Red-backed Squirrel Monkey is found in the scrub forest of Costa Rica while the Common Squirrel Monkey prefers the forests of Peru, Bolivia, Paraguay and Brazil. They are rather small monkeys, weighing no more than 1.1 kg (2½ lb) and they do not have a prehensile tail. When females are giving birth, the males are rather thin and tend to stay away from their mates but immediately before mating, the males put on a lot of weight and fight violently among themselves for the privilege of being a disinterested father a few months later. Often, a top ranking male is so exhausted after fighting with two or three other competing males — and, presumably, *after* claiming his breeding rights — that he has to rest for a few days before joining the group once more. The single young is born after a pregnancy of about six months and it has to be tough to survive, for while paternal support is non-existent, maternal care and attention is no more than minimal.

Cebus, or capuchin, monkeys are the only New World monkeys to spend much time on the ground. They are intelligent and will frequently use sticks in the wild to obtain food beyond their normal reach. They also open nuts by hitting them against branches and they spend many peaceful hours searching diligently for bugs under the bark of fallen trees and dead branches.

They are the traditional 'organ grinder' monkeys because they are small, easily trained and provide hours of entertainment with their lively antics.

Cebus monkeys can be divided into two groups. Two species of capuchin have tufts of hair over their eyes, or ridges of hair on the tops of their heads. These species are a uniform grey-brown colour. The other two species do not have these tufts of hair, have a whitish face, throat and chest while their limbs and the remainder of their bodies are predominantly black or at least very dark.

ABOVE
Almost uncertain of where to take her youngster next, an adult female Red Spider Monkey is caught in a moment of indecision by a jungle photographer. Her firmly attached prehensile tail is supporting virtually all of their combined weight which may be as much as 7 kg (15½ lb). Instinctively the youngster, oblivious to where it is being taken, is using its own prehensile tail which is wrapped around that of its mother, as a vital means of support. Hands and feet also secure a firm grip on her fur and by such means the youngster will learn the arts of treetop safety long before its ten month old dependency upon her comes to a reluctant end. Spider monkeys feed on fruits and nuts which are largely out of reach to other arboreal monkeys and, contrary to local belief, only very rarely — when their own food supply is desperately short — do they descend to the ground and damage cultivated crops, a habit for which they are often persecuted.

The woolly monkeys, after the howlers, are the second largest New World monkeys. They also have incredibly long prehensile tails which give them certain arboreal advantages over many of their relatives. One of the two species is fairly common throughout the forested Amazon basin while the other is found only on the eastern slope of the Cordillera Central in Peru. Ranging through mountain regions to a height of at least 2000 m (6600 ft), their name is derived from the appearance of their short, thick fur, especially on their head. They are diurnal and may move around in groups of at least 50 individuals, some of whom may be down on the ground while others forage in the trees. They walk, like the gibbons, on two legs using their outstretched arms to keep their balance. Occasionally, they will sit back and use their strong tails as props. They appear to feed almost exclusively on fruit and leaves and also to have a rather short pregnancy of between four and five months.

Closely related, but in name only, is the Woolly Spider Monkey from Brazil. It is probably one of the least well known of all primates and little is known of its habits other than that it is arboreal, diurnal and feeds on fruit and leaves. It has a prehensile tail and probably moves around in small groups.

Last but by no means least in the cebid family are the spider monkeys. They are the outstanding tree specialists of the New World and, among all primates, are exceeded in speed and agility only by the gibbons. Their limbs, including their prehensile tail, are incredibly long and they seem to glide across the tops of the branches. When alarmed, they will break off quite large branches and attempt to

drop them onto an intruder, and they will also emit barking and whining calls of disapproval. A normal group consists of up to 25 individuals, often in smaller groups which stay near each other as they feed.

The second family of New World monkeys contains the marmosets and the tamarins, and they are set apart from the rest because their hands and feet are equipped with claws instead of nails, although their big toe does possess a nail and the so-called claws are sufficiently flattened for them to be distinct from true claw-bearing animals. Their noses are dry and they do not have a prehensile tail. They are strictly diurnal, arboreal and they feed largely on fruit, leaves and on insects. They are among the smallest of the primates.

Marmosets are probably of fairly recent evolutionary origin and they are known mostly from the tropical forests of the Amazon basin, although they extend as far north as Panama in Central America and one lives in the tall grass of the Matto Grosso. Their thumbs have not evolved as opposable digits and their forelimbs are considerably shorter than their hindlimbs. When they feed and rest in the trees, they often use their sharp claws as hooks which are capable of supporting their body weight when they relax.

From one to three young may be born after a pregnancy of five months and the male plays a vital role in child-birth and in its subsequent care.

Interestingly, but probably not too significantly, these diminutive primates share important characteristics with Man; the role of the father in child-care is the obvious one but we also both possess 32 teeth and 46 chromosomes.

ABOVE
A captive Dusky Titi crouches on a branch in typical fashion, about to use the most efficient means of escape in the wild. Such a position means that the powerful hindlimbs are flexed ready to launch the monkey into mid-air at the slightest hint of danger. Dusky Titis travel in small family groups and scent-mark branches to convey information to others of their species. They usually live in small but well defined territories, and engage in vocal battles with nearby groups to establish and to maintain their rights in the forests. In direct conflict with each other, males raise their hair and increase their size in a threatening gesture. This species is more or less confined to the forests along the south bank of the Amazon River where it feeds on fruit, insects and birds and their eggs. Two other species in the titi group are widely distributed through South America's tropical forests.

The Pale-faced Saki is set apart from the Monk Saki in having a distinct white fringe of fur around its face. This species is completely arboreal but may come down to the lower branches of a tree when food is scarce higher up. If disturbed it will run quickly on all fours along a branch, but it will occasionally stand upright, waving its arms to keep its balance, and run expertly on its hindlegs — perhaps even to the end of a branch and then launch itself out into mid-air from tree to tree. Unlike many monkeys, which sleep in a fairly erect position, the Pale-headed Saki curls up in a cat-like fashion on a branch.

RIGHT
A rugged-looking group of White-fronted Capuchins are of the untufted type and have the characteristic dark arms and legs, the rest of the body being much lighter. Although they look a bit hostile in this picture, they are very lively creatures and are one of the most popular pet monkeys in the United States. Immediately after birth, the youngster hangs onto its mother as tightly as possible and as they grow they soon learn to ride on her back. Should it become separated from her at any stage during its dependency, it will cry out plaintively often inducing others to come to its aid. Capuchins have prehensile tails but they lack the bare under-surface at the tip and are therefore not as advanced as those of some other New World monkeys. Like the Chimpanzee, the capuchins have been found to use tools in the wild, especially twigs for probing for larvae in tree-holes.

LEFT
The soft and finely-graded mottling on a Pygmy Marmoset gives its coat the appearance of the feathers of a bird. This is the smallest of all the marmosets and it is found in the forests of Peru, Ecuador and Brazil where it is difficult to locate because of its small size, its uniformly brown appearance at a distance, and its agility in the trees. Typical of all the marmosets, this species has a lot of its face covered in fur which restricts the use of the whole face in communication, which is largely effected by movements of its lips at close quarters and by whistling calls at a distance. This latter habit is a highly efficient means of keeping members of foraging family groups in touch with each other, especially where visual contact is prohibited by dense vegetation. When alarmed, Pygmy Marmosets make their way up a tree in a spiralling fashion in the manner of some squirrels. They usually sleep in holes in trees during the night.

LEFT
The Cotton-headed or Cottontop Tamarin has earned itself this descriptive name because of the ample crest of white hairs that sprout from the top of its head. Over a predominantly white body it has outstanding, rusty coloured 'epaulettes' of hair which are probably used in displays. In threat displays, the head tuft is flattened and all the fur is fluffed out so that the size is visibly increased. Dominance between males is established by one male backing towards another with tail erected to expose genital scent glands. In such an encounter, the 'inferior' male acknowledges his status and retreats accordingly. A dominant male leads the group in search of food and the likelihood is that these tamarins are inclined to eat meat more than other species. Contact is maintained by high-pitched calls which may even be within the ultra-sonic range.

RIGHT
The Common Marmoset has a predominantly grey-brown body but outstandingly white tufts over its ears. It lives in the thick tropical forests of Brazil. The male is reputed to assist the female during labour after her five-month pregnancy, and after the young — often twins — are born he takes care of them but returns them to the female every few hours so they can be suckled. After three weeks the growing youngsters are strong enough to relinquish their hold on their parents' fur and to begin exploring for themselves, although they return frequently and are carried by their parents until they are about two months old.

LEFT
Perhaps the weirdest of all the primates, a hunched up Red Uakari presents the complete spectacle of dejection. But even with an infant clamped firmly to her back, a female Uakari is one of the best leapers of all the New World monkeys. She will rush at great speed along a branch at the very top of a tall swamp-bound tree and then sail out into mid-air with all her loose fur streaming out behind her. When she lands, she continues almost immediately with her slow and lethargic climbing that more befits her looks, leaving any onlookers almost misbelieving the awe-inspiring feat they have just witnessed.

ABOVE
The short hair of a Woolly Monkey gives it the appearance of being wrapped up to face the cold rather than the hot, humid jungles of the Amazon basin. In these forests, the Woolly Monkey is an expert arm-swinger, often stopping to scent-mark branches from special glands on its chest. It often moves to the edge of the forest and then drops vertically for as much as 10 m (33 ft) into the bush below to feed extensively on fruit and leaves. In fact it has been called *Barrigudos* by Brazilian natives in view of its voracious appetite. No doubt the natives approve of this gluttonous habit for they are rather partial to Woolly Monkeys themselves.

RIGHT
Resplendent with a stunning mane of brilliant orange, a Golden or Lion-headed Marmoset is one of the most beautifully coloured of all monkeys. Highly conspicuous to predators such as eagles, these arboreal primates have evolved the ability to move through the trees with tremendous speed until they have put a respectable distance between themselves and their hunter. Their nails bite deep into the bark of trees giving them an immediate grip and their movements are disconcertingly jerky. When they want to descend from the heights, Golden Marmosets apparently slide down tail first. They are restricted in their distribution to the forested mountains of south-eastern Brazil and their value to collectors has caused their numbers to decline in the past. Fortunately, it is now illegal to export them and they may still breed freely enough to gain in numbers once more.

MONKEYS FROM THE OLD WORLD

The wide-stretched mouth of a Hamadryas Baboon is not as it might seem, a yawn, but a fully alert threat display. This is the typical gesture of one male to another in conflicts over dominance, and it is one that all youngsters must learn to respect if the harmony of the social group is to be maintained. The Hamadryas is similar in most aspects of its behaviour to its Ethiopian neighbour, the Gelada, but it extends its range to Arabia, Egypt and the Sudan. Revered and often mummified by the ancient Egyptians, the Hamadryas figures prominently on many temple carvings. Its main predator today is the leopard and although adults will bark and scream at such an intruder, they are soon quelled if one of them falls prey to this animal.

ABOVE
Exploring the secrets of a Hindu temple in Katmandu, Nepal, Rhesus Monkeys enjoy freedom wherever Hindus worship Hanuman the monkey-god of healing and worship. Thus, they may be found in the busiest of cities as well as the wildest of jungles. In the Western world the Rhesus Monkey is found extensively in circuses and zoos for it is very hardy and thrives in captivity. Youngsters are ideal as household pets but there is a tendency for them to become bad tempered and aggressive as they grow older. Apart from its notable contribution to medical research, this was the first monkey to be successfully launched into space.

The second group of monkeys, those of the Old World, were very successful. They spread out through the great forests of Europe, Africa and Asia and were so adaptable that some of them were able to live on in areas from which the forests were shrinking. The dry and cold period of about 35 million years ago, which forced this reduction of the forests, created in their place many different sorts of habitat. In adapting to such places, the baboons and the macaques were forced to return to the ground. Thus, they largely relinquished the safety of the trees and as a result they developed powerful means of self-defence to keep hungry, terrestrial predators at bay. So, they became large and ferocious, roaming farther and wider as the forests, taking the majority of primates with them, retreated to the equatorial regions that they occupy today.

The Old World monkeys were stronger, bigger and far more adaptable than those from the New World and these characteristics may be significant in their being the forebears of the apes and eventually of Man. Fossil remains indicate how extensively they were once distributed across Europe and Asia and even today they occupy an incredibly large and varied area, extending through the African continent (except most of the Sahara Desert); from Gibraltar across to the Himalayas, South-East Asia and even to the snow bound regions of northern Japan.

The Old World monkeys are diurnal and are divided into two distinct groups. There are the guenons, the mangabeys, the baboons and the macaques all of which are predominantly omnivorous. They eat fruit, leaves and other parts of plants, but will readily supplement these with insects, small birds, mammals and lizards if they can catch them. All these monkeys spend some time on the ground and they possess special cheek pouches in which they store food as they are feeding — guaranteeing a useful reserve of food should they be put to flight unexpectedly.

The second group, consisting of the colobus monkeys and the langurs, are leaf-eaters. Such foodstuff, however, needs a lot of digesting for it consists mostly of cellulose which is hard to break down. But these monkeys have developed a stomach system similar to that of the cow to help them to digest as much as possible. They do not have cheek pouches, probably because they are predominantly arboreal, and leaves are abundant wherever they live. They are also subjected to fewer predators.

Characteristics of Old World monkeys are that they have nostrils which are close together, facing downwards and outwards, and their thumbs are fully opposable, although that of the colobus monkey is greatly reduced. Their tails are not prehensile and they have hard pads — known scientifically as ischial callosities — on their bottoms to act as cushions when they rest on branches or on the ground.

Among the cheek-pouch group, the guenons contain more species than any other Old World monkey genus. They are found only in Africa — south of the Sahara and north of the South African veldt land. The word 'guenon' is derived from the French language and implies 'fright' — a direct reference to the creature's habit of grimacing and exposing its teeth when it is excited or angry. Their displays also involve the use of the bright markings on their faces and bodies. These monkeys are active by day, especially in the morning and evening when the heat of the sun is not too intense.

The common social organization among Old World monkeys is a stable group containing several males and females, but there are three unrelated species which deviate from this plan. They are the Patas Monkey, the Gelada Baboon and the Hamadryas Baboon. All three live in open grasslands and woodlands where trees are scantily spaced or are too small to afford them protection against predators. Their social groups consist of one dominant male and several dependant females. Competition between males for this privileged status has led to them becoming almost twice the size of their females. They are large and powerful creatures and they have also developed, especially noticeable in the two species of baboons, mantles of flowing hair, powerful shoulders and enormous jaws equipped with ferocious-looking teeth. Such assets also serve to keep predators at bay, for an angered male is indeed a formidable sight.

The Patas Monkey is found in the savannah of northern equatorial Africa from Senegal to Kenya. It is not as well armed as the baboons and is more adapted for

rapid flight than for staunch resistance. Groups consist of up to fifteen, headed by a dominant male who maintains group contact by crashing through undergrowth and flashing his conspicuous white bottom to the females who may have strayed some distance away from him while they are feeding. The group may have a range extending over some 52 km² (20 sq mls) and the amount of it that they cover in a day depends solely on the amount of food they can find. During the heat of the day they rest in a group beneath the shade of a tree but at night time they sleep individually in separate trees — almost certainly an anti-predator device for a prowling leopard stands far less chance of catching them this way. They also change their sleeping sites each night.

Baboons are perhaps the best adapted of all the monkeys to a terrestrial life. They have adapted to habitats ranging from rainforest to semi-deserts; the West African forest species include the Mandrill, the Drill and the Guinea Baboon; the savannah species include the Chacma Baboon in southern Africa, the Yellow Baboon in the centre of the continent and the Olive Baboon in eastern Africa. These species may range far into open grasslands by day, returning to the safety of trees or protected cliffs at night. The Hamadryas Baboon is a semi-desert inhabitant of north-eastern Africa.

Baboons are mostly promiscuous, with one young being born to a female after a pregnancy of six or seven months. Two exceptions to this are the Hamadryas and

ABOVE
A young Gelada Baboon clings tightly to the fur on its mother's back as she forages for grass and feeds with other females in the rocky ravine of an Ethiopian mountain range. Very little is known about the breeding behaviour of these formidable, ground-dwelling primates, although it must be essentially similar to that of other baboons. Geladas spend nearly all of their lives on the ground and take to the trees only very rarely. They have few predators, although eagles will occasionally make off with a youngster who has strayed from the protection of its mother.

the Gelada Baboons both of which live in a harem system similar to that of the Patas Monkey. A noticeable feature of baboon life is grooming — either of themselves or of each other. Such activity helps to stabilize social groups with subordinate animals mostly grooming their superiors. The animal receiving such attention gives the impression of enjoying a state of ecstasy as it lies back, half closes its eyes and allows itself to be pulled and pushed around.

Somewhat smaller than the baboons are the macaques which are the only African monkeys to be found north of the Sahara. They haunt the rugged Atlas Mountains of north-western Africa but also extend their range eastwards through Asia from Afghanistan to Japan. A small but rather famous colony of these monkeys still lives on the Rock of Gibraltar where it is fed by the British army. Legend has it that should the Rock lose its macaques then the British will lose Gibraltar. Consequently, they are well looked after and Winston Churchill even ordered their numbers to be augmented from a North African population during the Second World War when they were on the decline. Today there are about 30 left and they are often called Barbary Apes because of their lack of a tail. They are, however, the only monkeys to live wild in Europe. Macaques live in troops of up to 24 individuals of both sexes and of all ages, although occasionally separate troops may join forces to form temporary groups of more than 100 animals. Like the baboons, they are strong and will offer stout resistance to potential predators. Macaques from Sulawesi (formerly Celebes), an island in Indonesia, have been known to kill wild dogs singlehanded.

A typical and well known species of macaque is the Rhesus Monkey. Hindus and Buddhists hold it as sacred, protecting and feeding it until it has become fearless — often to the extent of misbehaving itself. But its antics are tolerated by people who believe these monkeys to be reincarnations of their own ancestors.

Because Old World monkeys are biologically similar to Man, they are used extensively in medical and psychological research, and perhaps the most important discovery from the Rhesus Monkey has been of the Rh (rhesus) factor. This is a genetically determined protein in the blood and, in some circumstances, it can prove fatal to unborn and newly-born human infants. Its discovery has led to a cure being found and many human lives have been saved as a result of this.

Finally, in this group of Old World monkeys, are the mangabeys. They are quite small and are confined to the equatorial regions of Africa, including Guinea, Liberia, Uganda and Kenya. Very little is known about these monkeys in the wild although they share the common feature of having distinctive white eyelids which, when flickered, act as a means of visual communication between members of the same species.

The second group of Old World monkeys, the leaf-eaters, are easily divided into two geographical groups. They are the colobus monkeys from Africa and the langurs from Asia. The colobus monkeys, or guerezas as they are sometimes called, have long hairs on the sides of their bodies and tails which increase wind

LEFT
Facing the harsh realities of life in a cold climate, this family of Japanese Macaques, or Snow Monkeys as they are appropriately called, huddle together on a branch to keep warm. The Japanese Macaque is the most northerly-living monkey in the world and it is clearly well adapted to the cold climate by having a very thick, pale coloured coat. The snow may be 0.3 m (12 in) deep but they will wade through it in single file taking turns to lead the way. In some places they have taken to warming up in hot springs, and it must be an amusing sight to see them sitting in a row with their heads still covered in a layer of snow. The normal summer diet of these monkeys is succulent green vegetable matter, but when the winter snows descend they have to resort to feeding on seeds and pieces of bark taken from trees when they can find nothing else.

ABOVE LEFT
A picture of a Japanese Macaque taken during the summer shows her surrounded by an ample supply of vegetation to feed on, and lazily grooming her recently-born youngster. This is a stark contrast to the winter scene of freezing temperatures and hazardous survival. Mutual grooming in all animals is important for it not only rids them of external parasites but also helps in the formation and maintenance of friendly bonds. This devoted mother will search over most of her youngster's body, looking closely at the skin and hairs. Females do most of the grooming in the group, while the adult males relax and wait to be sought out by their subordinates.

ABOVE
A young Japanese Macaque has large and quite powerful feet to enable it to climb even though it is probably only about a year old. Born after a five to seven month pregnancy, to coincide with the melting of the harsh winter snows and the emergence of a plentiful food supply, this youngster was dependent upon its mother for at least twelve months until, the following spring, it took its first tentative steps away from her side. In this tree it looks rather helpless, but it will soon grow accustomed to its new-found freedom and begin to enjoy carefree days of exploration and tolerance by all in the group. But as the days go by it must be taught respect for its elders who will prepare it for later life by being less kindly disposed towards its youthful spirits.

ABOVE
High up in the forests of Sri Lanka (formerly Ceylon), a family party of Rhesus Monkeys engages itself lazily in mutual grooming. These monkeys rarely travel on their own, preferring the safety of up to twenty individuals, and they are equally at home in the trees and on the ground. In any group there is one dominant male whose confidence is conveyed to any potentially rival males in the way he moves around. His head is held high, his short tail carried dominantly erect and there is an atmosphere of supreme authority about every step he takes. The females produce one young after a pregnancy of between five and seven months at irregular times during the year, although the months of April and May seem to be the preferred time for birth. The offspring is hairless and weak, is nursed for at least a year and is not fully mature until it is about four years old, indicating that the Rhesus Monkey lives for perhaps as long as 30 years.

resistance and act as parachutes during their prodigious and spectacular leaps from tree to tree. They have reduced thumbs and long fingers which act as highly efficient hooks as they make their way through the trees. The state of their thumb has led to their particular name, for the Greek word *kolobus* means 'mutilated one' which is the appearance that their virtually four-digited hand gives. They live in social groups of about twenty and although a strict hierarchy does not exist, a strong male will cover the retreat of the group in times of danger.

The langurs of Asia are much less arboreal than the colobuses and, like them, show certain adaptations for brachiation — the art of moving through the trees, swinging from branch to branch with the arms. They also have a reduced thumb and a long tail which they use for balance.

While the breeding behaviour of colobuses is not well known, female langurs give birth to a single young after a pregnancy of seven months but, surprisingly, the mother adopts a rather carefree attitude towards her child. Occasionally, a jealous female will steal the infant but the deprived female puts up little resistance to this act, even if the female is from another group.

Perhaps the most extraordinary member of this leaf-eating group is the Proboscis Monkey which lives in mangrove swamps along rivers on the South-East Asian island of Borneo. The males have an enormously conspicuous nose which is probably associated with displays. It may hang down right over the mouth and can swell and turn red when the monkey is enraged or excited.

RIGHT
The very young Crab-eating Macaque, clinging tenaciously to its mother in a Hindu temple on the Indonesian island of Bali, is covered in dark hair which is quite different from the lighter grey-brown of the adult. The natives of Bali revere this monkey in the same way as the Indians revere the Hanuman Langur in India. They still perform elaborate ceremonial dances and even prepare offerings of food which are placed in the forests where the macaques will stand a good chance of finding them. Their natural diet includes crabs and other small animals, and they regularly go down onto mud flats at the ebb tide and wade around in shallow waters to hunt. They can also swim very well and will readily dive into deeper water after their prey. Like its relative the Rhesus Monkey, the Crab-eating Macaque has been used extensively in medical research and to it we owe the development of the polio vaccine.

The wise old man of the guenons, a De Brazza's Guenon is the most elaborately marked of his close relatives. Unlike many other monkeys, he makes very little use of facial expressions, relying entirely upon his ornate face colourings to convey his intentions to rival males and to potential mates. Guenons live in forests and savannahs throughout Africa south of the Sahara. The forest species, of which this is one, exhibit the most elaborate combinations of colour to make species identification a much easier task among the dimly-lit trees. A typical guenon party consists of one adult male, perhaps three adult females and as many as seven youngsters of various ages. The adult male will not tolerate the presence of another male in his group, so the maturing males over whom he presides have to become cautious as they approach the age at which he might consider their presence a threat to his status.

PREVIOUS PAGES
A baby Olive Baboon, clinging to its mother at an East African waterhole while an adult male relieves its thirst, is actually the centre of attention within the troop. Other baboons, especially females, like to groom it and its mother is often given a special status. Olive Baboons seem to enjoy a friendly and relaxed social structure, and the male hierarchy is not noticeably present although it certainly exists. This species sleeps in trees during the night when its only real predator is the leopard. Should such an animal be confronted by a troop during the day, it will soon learn that it is no match for several large males who defend their troop with incredible bravery. When the troop is on the move, young males walk at the edge and the older males accompany the females towards the centre. Should they all be threatened with danger, the mature males come forward to face the enemy while the remainder of the troop moves off.

ABOVE
Found only in the rocky mountains of Ethiopia above 1800 m (5905 ft), the male Gelada Baboon carries on its chest an intricate arrangement of pink, bare flesh and grey fur which combine in both sexual and threat displays. Females also possess a similar patch but they are themselves much smaller than the males and less densely furred. Geladas feed on roots, leaves, fruit and small animals and when wood is plentiful they will readily associate in very large groups, although when food is scarce they split up into smaller groups dominated by one male. Their warning cries are harsh barks and, at the sound of these, surrounding Geladas scamper uphill to the safety of caves and piles of protective rocks. If a predator pursues them too far they will resort to the trick of rolling rocks and stones downhill to discourage it.

RIGHT
The curious nose of an adult male Proboscis Monkey is by far the largest among all the primates and probably acts as an effective sound-booster when he calls out in a Bornean mangrove swamp. No doubt such a nose is also the result of selective pressure of breeding. Females, tending to be attracted to a male by the size of its nose, would produce equally well-adorned male offspring (the female nose is much smaller). These males would then compete among themselves for females and those with the largest noses would win out once more. Thus, after many generations their noses have become almost preposterously large but highly relevant to their daily lives. The Proboscis Monkey lives along river banks and estuaries on this single East Indies island, and feeds primarily on leaves and young mangrove shoots. Consistent with such a water-bound existence, these monkeys are excellent swimmers and will dive quite readily into water.

PREVIOUS PAGES
Experts in ground-dwelling life, four large Olive Baboons rest in the comparative safety of a bare-branched tree in East Africa. Baboons feed on almost anything they can find. So powerful have they become to combat the ground-dwelling predators which abound in Africa, that they are sufficiently fearless to attack even human beings. They will frequently attack domesticated animals and are among the great scourges of the tourist in Africa, jumping up on to vehicles and pulling eagerly at windscreen wipers.

LEFT
Held close within the safety of its mother's protective arms, a baby Chacma Baboon stands a far better chance of asserting itself over other baboons later on in life than does a youngster born to a less attentive female. From its mother it will learn much of the discipline and devotion vital to the successful social role it must play when it finally leaves her protection. Baboons have a well-defined social structure and even a large, scattered group feeding out in the open has sentinels posted around to keep a look out for predators. On the African savannahs predators are numerous and the baboons have themselves become fearsome primates to combat this threat. Their main danger is possibly the leopard which will attempt to catch them in trees where they sleep at night.

RIGHT
The Lion-tailed Macaque, looking like its carnivorous African namesake in its stance and pale face ruff, is a shy inhabitant of the dense forests of the west coast of southern India. This is essentially an arboreal monkey which moves through the trees in groups of between ten and twenty individuals. Usually a dominant male moves cautiously through the canopy some distance ahead of the main group. They feed on fruit, nuts, leaves and insects and often descend to the ground to play among themselves or to splash in water.

RIGHT
The Abyssinian, or Black and White, Colobus Monkey is probably the most arboreal of all of Africa's monkeys for it descends to the ground only rarely to lick earth for its vital mineral salt content. This operation does not take long and the rest of the time is spent feeding in forest or thick woodlands where it is found up to about 3000 m (9850 ft). Colobus monkeys gather in loose associations of up to twenty individuals and although there is no strict rank system, males tend to be the protective sex in times of danger. Their escape through the treetops is rapid and is effected by the lack of a thumb, elongated hook-like fingers and prodigious leaps from tree to tree. Even though they are very distinctly marked, they are very difficult to observe from ground level for their shape is broken up against the harsh light of the bright sky behind them.

Using a block of porous lava as a seat, a Vervet Monkey pauses to chew on a small grass-root it has pulled up in Tsavo West National Park in Kenya, East Africa. Predominantly a primate of woodlands and grasslands, the Vervet Monkey rarely strays far from water and, where food is plentiful, may forage in widely spaced out groups of up to 50 individuals. When alarmed, they utter short bark-like calls and scamper off on all fours with their tails held high over their backs. Their inquisitive nature does not allow them to go very far without turning round and sitting up on their haunches to inspect the intruder who has put them to flight. Should they feel too closely threatened, they will readily take to the nearest tree and even draw leafy branches around themselves to gain added protection.

A young De Brazza's Guenon, just able to support itself away from its mother, shows the wispy beginnings of the beard that will grow to characterize it later on in life. Twins may be born after the usual pregnancy of about seven months. The newborn young cling to their mother's fur and even entwine their tails with hers to give them added security as she moves around in the trees. When they finally leave her side, they play among themselves in the trees, but always under the close watch of the old, dominant male who keeps an eye open for such predators as leopards, snakes and birds of prey.

A fleet-footed Patas Monkey resting in Uganda, East Africa, is unique among the guenons for it has forsaken completely the safety of the trees. Its limbs have become long, slender and ideally suited for a speedy escape in times of danger. One individual has been recorded as travelling, on all fours, at 56 km/h (35 mph) alongside a car. Although the males have intricate displays by which they maintain group unity over quite large areas, such displays are also modified to announce the arrival of such predators as hyenas. A male will act as sentinel to a feeding party, either standing up on his hindlegs to peer over tall grass, or climbing into a tree to command a better view. When he senses danger he crouches down in the grass and utters a soft but penetrating call and when the others have taken note and crouched themselves, he begins a diversionary display to draw the predator's attention away from vulnerable females and youngsters. By moving off in the opposite direction the male lures the predator away and only when it has gone a safe distance will he return to his troop.

ABOVE
This picture of a Golden Langur clearly shows how langurs got their name. It comes from a Hindustani word *Lungoor* which means long-tailed and is highly appropriate for this monkey. The tail is not prehensile and is used primarily as a balancing agent for running along branches high off the ground. It probably gives added stability when jumping from tree to tree. Being almost exclusively arboreal like all langurs except the Hanuman Langur, the Golden Langur has a long, slender body with equally slender limbs. With such effectiveness does it live high in the treetops that it was not 'officially' discovered until as late as 1953 although its existence has been known since early this century. It is an interesting monkey for it undergoes seasonal changes of colour: during the hot summer months it is creamy white as shown here, but as the winter sets in it becomes much darker and seems to be covered in a golden wash.

ABOVE
The Capped Langur of Assam, Burma and Yunnan lives mainly in the dense rainforests of the Naga Hills and is a typical langur. It rarely comes to the ground and has the classic characteristics of a brachiator — a reduced thumb and long fingers and arms which are strong enough to support it as it swings from branch to branch. Langurs all have eyebrows of stout black hairs, but this species also has a peak of hair on the crown and white tufts covering its ears. These give it a rather striking appearance. Langurs live in groups of up to 40 individuals led by an adult male and they search for vegetable food during the day, probably resting during the heat of the day and returning to their sleeping trees as the dark of night closes in on them.

RIGHT
The Hanuman, or Common, Langur on the left of this picture shows the black, bristly eyebrows which are a feature common to all langurs. Like the Rhesus Monkey, these particular langurs are held as sacred animals in India and are found almost everywhere — from high evergreen forests to temple buildings in arid areas. This is the most terrestrial of all the langurs and also the largest — a fully grown male weighing up to 11 kg (24 lb). They live in two types of group, males and females together in which only one adult male is likely to preside over females and immature males, and just males on their own. They breed throughout the year and parental care is of the highest level of conscientiousness. When they are six months old, the young leave their mother's side and sub-adult males are expelled from the group by the dominant male when they are about nine months old.

ABOVE
Not much larger than a squirrel, this hand-held Talapoin Monkey is among the smallest of the Old World monkeys. Little is known about its behaviour in the wild except that it frequents thick forests in the Congo basin, extending eastwards to the Ruwenzori Mountains — the fabled Mountains of the Moon — on the Ugandan border. In such a place the Talapoin Monkey may range to an altitude of 2500 m (8200 ft). It is nearly always seen in close association with water and it likes to travel through the trees lining the banks of rivers and streams. Talapoins feed on fruits, leaves and insects and have even been recorded entering the normally hostile world of a native village in search of the abundantly grown maize. When alarmed they will produce a series of short, hissing notes.

THE LESSER APES

The astonishing length of the arms of a
White-handed Gibbon is shown, as an adult
rests on a rock surrounded by vegetation.
In such an attitude, the ungainly arms are
held out to clutch at the vegetation to give
the gibbon some support in this unusual
resting place. But up in the trees, where
gibbons spend most of their time, these
arms are superbly adapted and transform
this primate into one of the most graceful,
and probably the most agile, of all
mammals. Many of the primates propel
themselves from branch to branch by using
powerfully developed hindlegs but the
gibbons rely solely upon their arms to
swing their bodies through the trees. These
arms are so specialized that they are about
two and a half times the length of the
gibbon's body and about one and a half
times the length of its legs. So far no fossil
gibbons with such long arms have been
found, suggesting that this unique feature
may be of quite recent evolutionary origin.

When they stand erect in the trees and
support themselves with their hands, the
gibbons look uncannily human. Their faces
have a soft, almost weak, expression and
their large, warm eyes have an appealing
quality matched only by human sympathy.
In this photograph, taken in a rainforest in
Borneo, the gibbon's opposable big toe is
well demonstrated as it clings powerfully to
a smooth and steeply angled branch. Its
long arms enable it to swing through the
trees and to reach out for otherwise
inaccessible food. Gibbons have short
trunks but ample chests, and their rigid
backs are held straight as they move
through the trees, which explain why they
have such difficulty in walking on the
ground. The short, straight back which is
necessary for swinging from branch to
branch is fairly inflexible and prevents the
gibbon from bending over to walk on all
fours.

From the world of monkeys we now turn to the apes, some of which spend much
of their lives on the ground moving around comparatively slowly. One ape can
claim to be the most spectacular aerial acrobat of all the primates. This is the
gibbon, which is found in South-East Asia. The tailless gibbons are undoubtedly
apes, for all the apes by definition have no tails, and their life in the trees means
that they probably represent the earliest surviving evolutionary division
between the Old World monkeys and the apes. It should not seem too surprising,
given Nature's devious ways, that the world should be furnished with some
largely ground-dwelling monkeys — the macaques and the baboons — and some
highly specialized tree-dwelling apes — the gibbons and their close relative, the
Siamang, also from South-East Asia. Together, the gibbons and the Siamang are
known as the 'lesser' apes, into which category they are grouped because they
possess certain anatomical features which are shared by all the apes and which
are absent from the monkeys and the prosimians. For example, over and above
the lack of a tail, the basic arrangement of the apes' teeth is quite different from
that of the monkeys, and it would seem feasible to equate this with the change in
diet — which led to an emphasis on fruit-eating — that the earliest apes must
have undergone. Gibbons are also different from the other apes because they have
large horny pads on their buttocks. These pads serve an important function —
that of comfort — and they illustrate the difference of life-styles between the
gibbons and other apes.

The term 'lesser apes' by no means implies that the gibbons and the Siamang
are at all inferior. It is purely a label of convenience, implying that, while they are
certainly apes, they did not pursue the same evolutionary line that eventually
gave rise to Man, the 'highest' primate of them all.

The most noticeable feature of the gibbon family is their size compared to that
of other apes. A full-grown Gorilla may weigh as much as 159 kg–182 kg
(350 lb–400 lb) and stand a full 2 m (6 ft) at the shoulders, while gibbons never
weigh more than 13 kg (29 lb), usually from 5 kg (11 lb) to 8 kg (17.6 lb). They are,
indeed, miniature apes which remained smaller than many of the monkeys
because they never came down from the trees to do battle with large predators or
to compete with the other emerging apes.

But, ironically, the lesser apes were committed to evolving along many of the
same lines as the other apes and this meant that they had to forgo the tail that had
proved so indispensable to the monkeys! Without a prehensile tail of the New
World monkeys and without the stout balancing tail of the Old World monkeys
the gibbons gradually evolved longer and longer limbs, especially their arms.
Today, a gibbon standing in an upright position drags its hands on the ground in
an ungainly fashion. It seems as though these limbs have been stretched by
centuries of use high in the trees as the gibbons became more and more expert in
their way of life.

Gibbons travel more quickly than anyone who has not seen them can possibly
imagine — looping from branch to branch at breakneck speed or hurling
themselves across wide gaps between supports. The speed of their reactions is
nicely demonstrated by C. R. Carpenter, an American who spent most of his life
working on primates. Carpenter once saw a gibbon launching itself into mid-air
but, as it did so, the branch it was leaving snapped and the poor creature had no
hope of reaching the other side because all the power of its jump had been lost. But
instead of falling, the gibbon, in one continuous movement, was able to twist itself
in mid-air, grab hold of the remaining branch stump, haul itself back into an
upright position and then launch itself across the 9 m (30 ft) gap to the next tree!

Little wonder, then, that the scientific name for the gibbon family is
Hylobatidae, which is Greek for 'tree walker'. The name is actually derived from
the gibbon's less spectacular and more normal method of moving through the
trees — swinging itself from branch to branch in an easy and relaxed fashion.
Technically this means of locomotion is known as brachiation and it is ideal for
making a hurried retreat from predators or for reaching feeding spots prohibited
to less agile creatures. It is a feature shared by all the apes, but the Gorilla, the
Chimpanzee and the Orang-utan use it far less than their smaller relatives.
Interestingly, the South American spider monkeys also employ this manner of
locomotion, which gives them certain affinities with the lesser apes although they
evolved quite separately from them.

If we accept gibbons as being the most primitive of the apes, then the simplest

Occasionally a black adult will challenge
even a human intruder into his territory
and will present himself on two legs,
hooting and arm-waving in an attempt to
frighten the trespasser into retreat.
Ironically, much of this arm-waving is
performed so that the gibbon keeps its
balance, but it obviously plays an
important secondary role as a threatening
gesture and has evolved to look extremely
menacing. While it is frantically waving its
arms and hooting, the gibbon stamps its
feet defiantly and makes small threatening
gestures, pretending to rush at its
adversary. Its flashing white face and
formidable antics are usually enough to put
most intruders to flight. The difference in
length between its hind and forelimbs is
actually not as pronounced as one might
suppose from most photographs, for its
hindlimbs are invariably kept doubled up
to suit its basically tree-top life and it is
rarely seen, as here, on the ground.

The black male Concolor Gibbons
occasionally have white patches on their
cheeks but the spectacular females turn a
rich golden fawn colour which, combined
with a plentiful abundance of shaggy fur,
sets them off magnificently against their
sombre forest backgrounds. As with all
species of gibbon, males and females mate
for life, for once the vital relationship so
important for protecting the slowly
developing young has been formed, it is
almost impossible for it to be broken. Some
captive gibbons have been known to live for
more than 30 years.

interpretation is that they evolved from more primitive, gibbon-like ancestors
which eventually gave rise to Man as well. This idea neatly groups the gibbons
with Man along the same evolutionary path, although we should always be
prepared to adapt our views in the light of fossil evidence.

The earliest gibbons almost certainly lived, as did the earliest Old World
monkeys, in Africa, where fossils dating back as much as 22 million years have
been found in Uganda and Kenya. The point about these fossils is that while they
are extremely gibbon-like in their skulls and in their teeth, the rest of their
skeletons are quite different. Their arms and legs are of almost equal length,
suggesting that they used all four limbs equally as they clambered around the
trees. Further evidence suggests that the gibbons may have evolved from these
ancestral quadrupeds after the main ape group had split away from the Old World
monkeys.

In all there are about six species of gibbon and they exhibit such individual
variation in their colours — depending on which species they belong to, or their
age and sex — that unless they present themselves in a threat display it is
virtually impossible to identify them reliably from quick forest sightings.

The gibbons' means of locomotion carries them swiftly through the trees and
they may cover large distances through the forests. High up in the trees, their
buttock pads provide them with an ideal alternative to nests for they simply sleep
on a branch, using their well-protected backsides as cushions. By comparison the
great apes move relatively slowly by day and when they come to rest each evening
they build nests of vegetation in which to spend the night. Such nests may only be
simply constructed, but they represent the first primate steps that finally led to
Man and his house-building instincts.

Gibbons are found over a wide area of South-East Asia, ranging from sea level
to about 2400 m (8000 ft), where they inhabit the thick tropical forests. They must
be among the noisiest of mammals, especially in the early morning when
neighbouring families remind each other of the borders of their territories.
Typically, a family consists of an adult male and female accompanied by their
youngsters. As the first light of a tropical dawn filters through the trees, the
female stirs her limbs and begins to emit her penetrating hooting calls. She starts
off slowly but rises gradually to a crescendo of piercing calls that ring out through
the awakening jungle. And then silence before the echo-like reply from the female
of a neighbouring family. Verbal contact established, the two families move
quickly through the trees until they make visual contact with each other. At this
point the males take over. Urged on by the hooting females, they rush at each
other through the trees, swinging and leaping to and fro, never crossing the edge

ABOVE
This gibbon, although it is probably the same species as those already illustrated, shows how remarkably individuals may vary in their body colouring. Usually in the animal world, such an extreme would be singled out by predators which would notice it more easily than its darker and better camouflaged relatives. By attacking these unusual individuals more often, the predators would prevent them from breeding and from producing similarly coloured offspring and eventually the odd colour would disappear. But the case of the gibbons is slightly different because they change colour as they grow older. An aged black male may very well, as a baby, have had a first coat of almost pure white hair, but while it was so vulnerably coloured it was constantly in the protection of its ever-alert parents. The only real threat to the gibbon is the leopard and even such a stealthy animal as this would have the utmost difficulty in approaching too closely, especially as gibbons sleep in such inaccessible places at the tops of trees. In the northernmost part of the gibbons' range, there are two species whose body colouring is related directly to whether they are male or female. In both species all the young are born pale, but as they mature they turn dark so that at about five or six years old they are all dark coloured. From then onwards, the males remain dark, eventually turning almost black, while the females become increasingly brown.

In close up, the face of a Siamang is superficially similar to that of a Chimpanzee. It seems to have a far more protruding bottom half to its face than the rest of the gibbons, but in fact this feature is accentuated by two points. There is no distinctive white border to the face to highlight the whole area, and the lower jaw and the area around the mouth are covered in prominent white hairs which draw attention to this part of the face. The Siamang is just as much at home in the treetops as all the gibbons and, like all the apes, it carries objects with its feet while it swings from branch to branch, although it may be slightly more limited in its carrying capacity because of the strange feature of its second and third toes being united by a web of skin.

In many respects the behaviour of the Siamang is typical of the gibbon family, but the female Siamang is pregnant for a few weeks longer than average and Siamangs are predominantly leaf-eating rather than fruit-eating. This is just as well in the areas where the Siamang occurs along with other gibbons, because it means that they do not compete for the same food.

OPPOSITE
Perhaps one of the most striking differences between the Siamang and other gibbons is the bare throat of the Siamang. This adult female, playing in captivity with her youngster, displays this fact superbly. She is actually calling, and in so doing she is distending her throat which is specially designed with loose skin so that it can be swollen with air to form a vocal sac at least the size of her head. If she draws air in through her nose when her mouth is closed, she can produce a deep booming note. When this air is expelled, a higher and more devastating note is produced. These calls are therefore produced alternately and co-ordinate quite simply with the Siamang's breathing. The speed at which they are produced varies according to the situation for which they are required. The Siamang can quite justifiably be regarded as the noisiest of all the noisy gibbon family, its piercing booms and howls echoing over distances of several kilometres shortly after dawn each day.

of the adjoining territories and always maintaining a respectable distance from each other. After a while this activity dies down, and the families move peacefully away from each other to commence their daily routines of grooming and feeding.

Such potentially aggressive behaviour seems to be specifically concerned with establishing a claim to each other's territory, with no actual harm intended to befall the individuals involved. Gibbons feed extensively on fruit, and as the food-bearing trees ripen at different times of the year the succulently laden trees are indeed prizes worth retaining. Each morning the contact calls ring out and the mock battles take place so that all the gibbons know the exact whereabouts of their nearest rivals.

The family unit of gibbons is strongly preserved because the young take at least six years to mature. The adult female gives birth after a lengthy pregnancy of just over 200 days, and the virtually hairless and helpless infant clings tenaciously to her chest, even while she swings effortlessly through the trees. When she is resting, the mother raises her knees to form a snug and furry cradle in which the youngster sleeps peacefully.

By the time the young gibbon is about two years old it is fairly independent, feeding itself and moving freely through the trees on its own. It is, however, still an adolescent and must remain within the family unit for a few more years. Freed from the immediate physical responsibilities of child-care, the mother — who has already been pregnant for several months — gives birth once again and, similarly, when this next baby is finally released from her care, she gives birth once more. Thus, the family unit grows in two-yearly stages until the first-born is fully mature and its presence is gradually resented more and more by its parents who will not tolerate other adults within their territory. The day finally dawns when minor squabbling turns to aggression and the scene is set for the six to ten-year-old to move away into the forest to fend for itself while it searches the trees for a potential mate in a similar predicament.

The Siamang, as we have seen, is closely related to the gibbons. It lives in the same area although it is more restricted in its distribution, occurring only in parts of mainland Malaya and on the island of Sumatra. There is a smaller kind of Siamang which lives on some of the smaller islands close to Sumatra.

The Siamang lives in thick rainforests, but it is found over slightly different altitudes from its relatives — mostly between 600 m (1975 ft) and 2400 m (8000 ft). It is slightly heavier than the gibbons, weighing anything from 8 kg (17½ lb) to 13 kg (28½ lb), the heaviest weight recorded for any member of the family. Its arms, compared to the length of its legs, are even longer than those of the gibbons.

THE GREAT APES

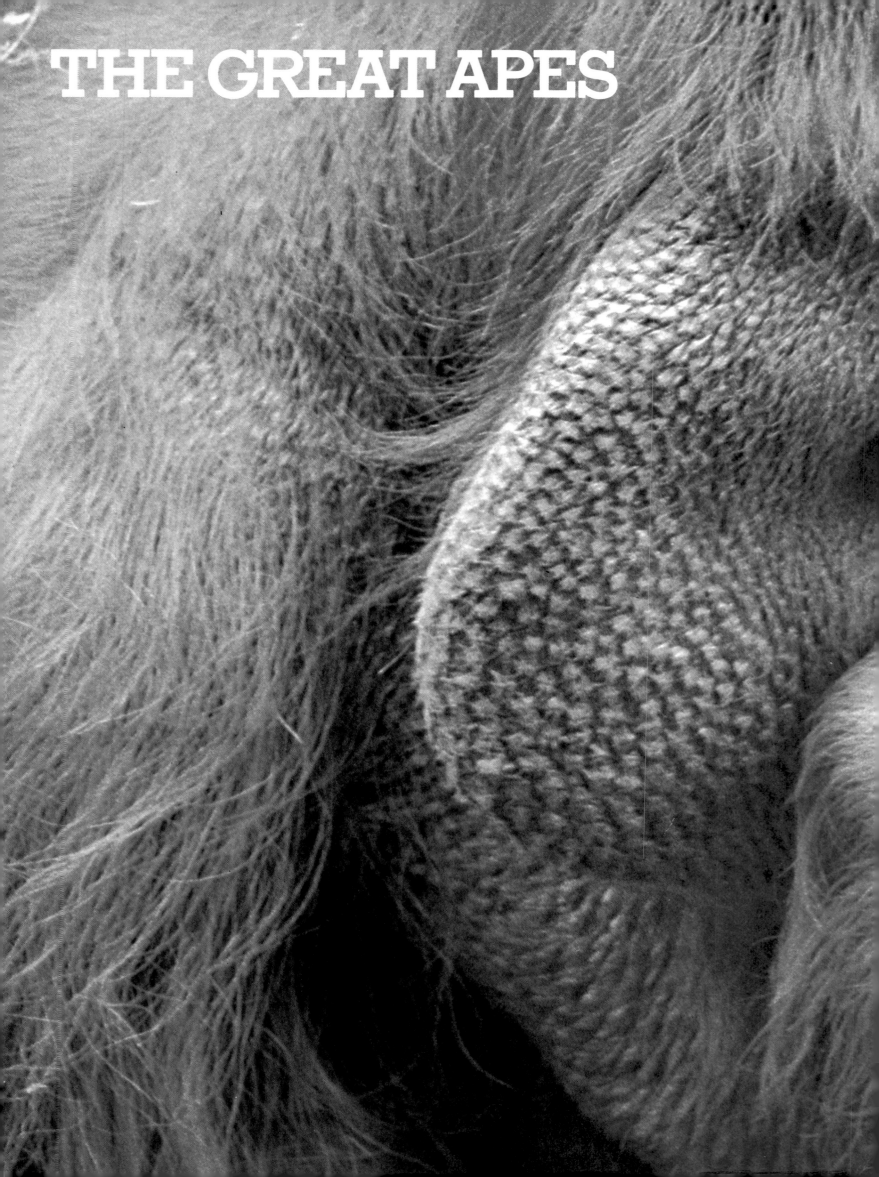

In close up, the male Orang's face is an extraordinary arrangement of fleshy 'blinkers', protruding lips and diminutive eyes which have certainly shrunk in size since the earlier examples of bulging prosimian orbs. The Orang is a day-living creature and has developed these intricate characteristics for several reasons. Threat displays between males involve visual contact and the more exaggerated the ornaments, the more awe-inspiring they will appear to other rivals. Females are more likely to yield to the biggest and the best-looking males who in turn will retain a stronger hold over their females. The 'blinkers' are deposits of fat which are not only important for display purposes but also contain valuable reserves of fat which may be drawn upon during lean times of fruit supply. Of all the forest offerings, the Orang seems to have a predilection for the tough fruit of the Durian tree. Such a fruit may be up to 0.3 m (12 in) long but the apes' protruding lips harbour teeth capable of tearing their way through the prickly covering to expose the fleshy and succulent contents within.

The adult male Orang is indeed a formidable creature. Ugly, if not even grotesque by human standards, his massively bulky frame is plentifully endowed with folds of burdening skin, protruding bulges of fat and muscle, most of which is covered by a coat of shaggy, tumbling, orange-red hair. But the life he leads seems to suit this display of threadbare obesity. He can have no natural predators and his dominance in the trees is rivalled only by other males of his kind. But such a frame can be moved only slowly and cautiously through the forest trees and the male Orang, referred to as the patriarch, for he may exercise control over more than one female in his territory, covers quite a large area in search of the fruiting trees that ripen at different times of the tropical year. Living himself to possibly the great age of 40 years, the male acquires vast and valuable knowledge as to the whereabouts of such foodstores, and the child-preoccupied females rely to a large extent on his guidance from one site to the next.

It is little more than 100 years since Man was startled by the revolutionary idea of a process of evolution put forward by such original thinkers as Charles Darwin and Alfred Wallace. Both men, but particularly Darwin, scoured the world amassing observations on living creatures, fossils and rock formations. Independently and almost simultaneously, Darwin and Wallace hit on the idea that no plant or animal had been created spontaneously, out of nothing, to exist for ever in that particular shape and form. Far more likely, according to their thinking, was that life began in a simple form and progressed from there, adapting to new environments and becoming more advanced with time.

They realized, of course, that for all of this to happen the world would have had to be millions and millions of years old by the middle of the nineteenth century, and this was something that they just could not prove. The experts at the time, believing that the sun was slowly burning itself out, could agree on no more than a maximum of 20 million years. But, as we now know, time was indeed on Darwin's side and the world is at least 4600 million years old.

The second question was that although Darwin understood that characteristics were passed from parent to young, he had no real idea of how such a process was carried out — at least not to the extent of being able to satisfy his critics. But what about these critics? Obviously such an overwhelming idea was bound to cause a stir among scientific circles and arguments for and against it would abound. But there was something about it that set the whole of Victorian England on fire. The one inescapable fact of 'Darwinism' was that Man himself was included in this great new theory of evolution: that Man, like everything else, was a product of the evolutionary tree. Scandal indeed! Was the Book of Genesis to be abandoned, challenged by a handful of madmen? Charles Darwin must have been terrified by the implications of his investigations and he even delayed the announcement of his theory — supposedly on scientific grounds — for many years. He was spurred into action by Alfred Wallace who quite independently had happened upon the same revolutionary idea.

Darwin's greatest apprehension may have stemmed from the fact that he came from a highly religious family. Indeed he himself was bound for a life in the Church until he seized the opportunity — despite paternal opposition — to join H.M.S. *Beagle* as a naturalist on its five year journey around the world when he left Cambridge University at the age of 22. He set sail in 1831 under the command of Captain Fitzroy who, perhaps, was to feel largely responsible for the devastating theory to emerge later.

But such were the moral problems confronting Darwin as he pieced together his theory of life that he was meticulously careful never once to state the obvious fact about Man's evolution. Instead, he suggested that we evolved from a 'hairy, tailed quadruped, probably arboreal in habit'. The implications were clear enough but Darwin played down the whole idea, ignoring it almost completely in his most famous book, *On the Origin of Species,* published in 1859. He concluded his 1871 book *The Descent of Man* with the almost apologetic idea that if we *are* part of such an evolutionary process then we may, at some future moment in time, expect to rise to a position even higher than that we occupy today. But what about these Great Apes that the world, only 100 years ago, could not accept as being our ancestors? They are animals that we all know very well. Man apart, they include the Orang-utan from the deep jungles of the islands of Borneo and Sumatra in South-east Asia, the Chimpanzee from of equatorial Africa and the Gorilla, also from equatorial Africa but restricted to the lowlands in the west and the more mountainous regions near Lake Victoria in central Africa.

The Great Apes appear to have evolved from an Old World monkey ancestry — which would help explain their absence from South and Central America today — at least 25 million years ago, although such an arbitrary date can represent no more than our interpretation of what may have happened. What seems clear, from fossil records, is that about 20 million years ago, during the Miocene era, there were at least four species of apes roaming the plains of Africa, and that before that apes were more extensively distributed across Africa, Europe and Asia. Our interest in them stems largely from our attempts to understand the immediate evolution of Man to whom they are all closely related. While this complicated step was being undertaken some of the apes were restricted within a forest existence where they remain today. Of all the primates they are the closest

An adolescent female displays the remarkable elasticity of the Orang's limbs. A true contortionist of the trees, the Orang is able to clamber, swing and climb its way through vines and branches to reach food in the most inaccessible-looking places. These flexible youngsters seem ideally constructed to indulge in the art of brachiation — arm swinging from branch to branch — and yet they lose this as they mature, especially if they are male, and their body weight increases too much. Indeed, study of Orang skeletons in museum collections has revealed a high incidence of healed bone fractures which suggests that branches frequently break under their weight. Such unfortunate individuals almost certainly owe their lives to the very denseness of the secondary vegetation of the forest which helps to break their falls.

to Man. They resemble him in many anatomical features even though outwardly they have developed many superficial differences — they are mostly covered in hair, they live in trees and when they walk on the ground it is usually on all fours.

The Orang and the Gorilla are large and heavy, moving with a deliberate slowness which befits their rather introverted life-styles. Chimpanzees, on the other hand, are lighter by comparison and are far more excitable and are extrovert by nature. This latter quality has led many people to believe that Chimpanzees are far more intelligent than the other non-human primates, and yet such a conclusion may only be drawn from the fact that the Chimpanzee's character permits it to explore and respond to experimental conditions with an eagerness not readily shared by Orangs or Gorillas.

But certainly it is to the Chimpanzee that most humans are attracted, for Chimpanzees revel in almost-human situations and as a result of this provide endless hours of entertainment at the Zoo. Gorillas and Orangs in captivity, by comparison, seem to be objects of pity for they are very inactive and take on a gloomy appearance as they sit virtually motionless in their cages.

Perhaps the main point about these apes is that they have large and complicated brains, although they are still one third the size of that of Man. This corresponds with their living for anything between 30 and 50 years and consequently the young taking several years to mature. As a result, protective family patterns are very pronounced. Usually one male reigns over more than one female and several immatures in various stages of development. All of them cease their daily wanderings in search of food shortly before sunset to build crude but effective nests out of pieces of vegetation. They build a fresh nest each evening.

Orangs and Gorillas are predominantly vegetarian although occasionally they mix their diet with some animal food. Chimpanzees, on the other hand, are renowned for their carnivorous tendencies and they will readily supplement their more regular fruit with termites, monkeys, pigs and other manageable forest creatures. There is even a record of a group of Chimpanzees attacking a native

The female Orang is much smaller and
more daintily assembled than her
masculine mate. As she clasps her infant in
her hidden left arm, the size of her right
arm indicates the tree-swinging power
provided by her enormous forelimbs. The
twig-sucking youngster grips her leg with
its foot, maintaining a strong hold which
suggests not only its dependency upon its
mother but also its instinctive awareness of
the need to remain firmly attached should
she suddenly take fright and make for the
treetops.

Although female Orangs are rarely seen at
ground-level, especially when they have
young in close attendance, this female is
the captive Joan who mated successfully
with a wild male in the Sepilok Reserve in
North Borneo in 1967. This was a major
triumph, for not only are Orangs
susceptible to many human ailments but
they are also prone to suffer from
malnutrition as a result of the difficulties of
providing a naturally balanced diet.

ABOVE and LEFT
In a rare moment of being photographed
away from its mother, a baby Orang learns
the first processes of the art of feeding itself.
Here it is using all of its dextrous limbs to
explore and chew a leaf-bearing twig
broken from a nearby tree. Normally, the
baby Orang remains under the close
protection of its mother until it is weaned at
about four years of age. During this time it
is cared for with a devotion paralleled only
by human mothers. The baby Orang is born
so helpless that its mother — the male
plays no part in child-rearing — must teach
it literally everything it needs to know to
survive. She will present it with
pre-chewed food, encourage it to clamber
alone among the branches of the trees and
even take it occasionally down to the
ground in an attempt not only to teach it to
walk but also to familiarize it with all
levels of life in what will always be a hostile
forest.

RIGHT
Too large and too energetic to be carried underneath its mother, a young Chimpanzee of at least six months is promoted to riding on her back as she moves around the open savannah. At about this age, the youngster grows on its tail a white tuft of fur which acts, quite literally, as a flag of truce when the infant goes exploring among the troop. At such a tender age the Chimpanzee can be of no sexual or dominance threat to older Chimpanzees and his presence is tolerated. So, while this tuft remains, the youngster enjoys family lenience but as it grows out so the attitudes of the elders change. The maturing animal must now learn the laws of his family group and strict discipline.

ABOVE
The epitome of peace, a young Chimpanzee rests its head against a rock to sleep. The Chimpanzee is almost certainly Man's closest relative, and, apart from the hairiness, the similarities can be seen in this face. Of all the non-human primates, the arrangement of body hair is actually the most human-like among Chimpanzees, the obvious differences being that theirs grows far more strongly in places where ours is greatly reduced. Some Chimpanzees, especially the females, are prone to baldness, and then the human-like quality of the shape of the skull is also apparent.

woman and killing her child but whether their intention was to devour the small human was left unproven.

The apes have developed the art of vocal communication to quite an intricate degree — although vision and hearing are probably more advanced. They may cover large areas in thick forests and it is important for them to maintain group contact as they forage separately for food. Their repertoires include hoots, growls, roars and even foot-stamping and tree-bashing to indicate their feelings towards each other and potential enemies. Of the latter, they can have virtually none in the wild — perhaps the odd leopard will take one here and there; but their most dangerous enemy is Man himself as he makes inroads into their habitats, or shoots them for sport or out of ignorance of their peace-loving natures. Fortunately, much of this thoughtless behaviour has been curtailed today, although a lot of trapping for export continues, and Man must still be considered a major threat to their future existence.

Sexual dimorphism — the difference in appearance between males and females of the same species — is only slight among Chimpanzees, pronounced among Gorillas and is highly exaggerated among the Orangs. A clue to the reason for this discrepancy may be that while male Chimpanzees and Gorillas are closely associated with their females and young, the male orang is a solitary creature. He has no immediate hold over his females who will respond to the fittest-looking males, who are therefore competing among themselves for the females. When a female mates with a chosen male, perhaps lured into his domain by his magnificent appearance, his special mate-winning characteristics will be inherited by any male offspring she should bear. Thus, over the generations, there is a trend towards males becoming more and more marked in their physical appearance. Gorillas and Chimpanzees have not reached these proportions because the males spend much of their time with their prospective mates and when the time comes for child-bearing, the breeding rights have already been established largely by the nature of the group structure.

ABOVE
Splashing around in the shallows of a slow
moving African river, a Chimpanzee
reveals an affinity with water not readily
shared by other apes. The apes often live far
away from rivers and while the Gorilla has
never been observed to drink, it
undoubtedly derives enough moisture from
its succulent diet. Jane Goodall, a pioneer
in the field of Chimpanzee study, reported
them as mashing leaves up in their mouths
and later using them as sponges to soak up
drinking water from nooks and crannies
that they could not reach with their
mouths. Chimpanzees are in fact well
known as tool-users and they commonly
insert strips of prepared grass into termite
nests. When the termites climb onto the
invading foreign body, the Chimpanzee
hauls it out and devours them.

RIGHT
In a position of troublefree, gay abandon an
infant Chimpanzee is subjected to a routine
body inspection by its mother. Born after a
pregnancy of between 200 and 260 days,
the helpless youngster weighs around 2 kg
(4½ lb) and is held fast by its mother,
cradled by arms and thighs in easy reach of
her breasts where it feeds at frequent
intervals. The adult female gives birth once
every three to five years and, not
surprisingly, the occasion is of special
interest to other members of the
Chimpanzee troop — in particular the
other females who are keen to inspect the
new arrival and to hold it themselves if
allowed to do so. When it is a few days old,
the infant is strong enough to grip its
mother's fur while she moves around in
search of food.

ABOVE
Caught red-handed, an adult female Chimpanzee, loaded to capacity, stares guiltily towards the camera of what must be a rather amused photographer. But this endearing habit can cause problems for the grapefruit farmer, for the Chimpanzees are delightful creatures to have around and yet once they realize that food is available in plenty, they can make substantial inroads into succulent crops.

This female shows quite clearly the swellings which not only identify her sex and maturity to the observer, but also indicate to the male Chimpanzees that she is ready for mating — even though she is prepared to cooperate with their desires for only a short period while she exhibits these outward signs of apparent willingness.

LEFT
In a West African rehabilitation centre, a sub-adult Chimpanzee demonstrates the ease with which members of his species take to the trees. Chimpanzees are, in fact, highly arboreal animals and they probably spend as much as three quarters of their lives up in the trees. Because of their light weight in comparison to their body size, Chimpanzees are well adapted for brachiation and they use this form of locomotion quite frequently over short distances. However, the arrangement of their thumb makes it difficult for them, as it is better adapted for hooking onto branchcs than it is for grasping them to guarantee a vital hold high above the ground. When on the ground, they normally travel on all fours, employing the knuckle-walking method of the Gorilla. In long grass they will walk quite happily in an erect or semi-erect posture.

RIGHT
The male Gorilla must rank as one of the most magnificent of all animals. In stature it dwarfs even Man, with its vast muscles and commanding frame. A confrontation with a male Gorilla is an experience that nobody forgets, for they have come face to face with one of nature's truly gentle giants.

Today we owe our appreciation of these shy and peace-loving forest inhabitants to people like George Schaller and his wife Kay who cast aside their fears and their firearms and spent many patient months studying the Mountain Gorilla in Zaire. They learnt their language, lived their life and always presented themselves as friends rather than as enemies until the gorillas grew accustomed to their presence. The result of their dedication has been a completely new appraisal of this declining, amiable giant of equatorial Africa.

LEFT
Salome is the first Gorilla to have been born at London Zoo. She received much publicity and caused quite a scientific stir when, on 16 July 1976, she made her appearance in the heart of London. Unfortunately, her father is not the famous Guy who refused to cooperate when presented with her mother-to-be for a few years. But all was not lost because the Bristol Zoo provided a willing mate. Salome's progress is being carefully monitored and she is receiving much human attention during her formative months in her concrete jungle. Dr. Robert Martin of the Wellcome Institute of Comparative Physiology in London Zoo has a son of a similar age and the parallel development of the two infants is being watched hopefully to throw a better light on our understanding of the evolution of 'higher' primates.

LEFT
Well hidden among dense vegetation, a male Mountain Gorilla peers timidly but suspiciously from under cover. His silver-grey back is just visible, indicating that he is a mature adult. Only when he senses that the cameraman has approached too closely will he perform his intimidating repertoire to maintain the safety of his jungle stronghold. He will issue grunts and snarls of disapproval, move around in an agitated fashion, shaking branches before making threatening gestures. He does not want any physical contact with his adversary but if his initial warnings are not heeded he may rise briefly to his feet, gathering himself to his full height, swelling his enormous chest with air and beating it with cupped hands — not with clenched fists as is widely supposed. With the echoes of this self-inflicted onslaught still ringing through the forest, he drops to all fours and rushes around in a frenzy, tearing up vegetation and swiping at anything that stands in his way.

ABOVE
A Mountain Gorilla climbing a tree must indeed be considered a rare sight, although their lowland relative will often do so to a height of about 5 m (16 ft). It is not that the Mountain subspecies is anatomically ill-suited to such antics but more likely because high altitudes do not normally support trees strong enough to bear the weight of a primate that may weigh as much as 273 kg (600 lb). On the ground, the gorilla progresses on all fours and is referred to technically as a 'knuckle-walker'. The last joints of its fingers are well padded and are covered in tough skin to accommodate what must seem to humans an agonizing means of moving around.

RIGHT
Basking in the Spanish sunshine, Snowflake, the only albino Gorilla ever recorded, is the object of much curiosity and excitement in Barcelona Zoo where he has been living since his mother was shot in the wilds of lowland West Africa for raiding a banana plantation. Snowflake — his Spanish name is *Copita de Nieve* — was removed to Europe where human foster parents nurtured him through his early vulnerable years. Once he had matured, he mated successfully with two normally coloured females, producing a daughter and a son. Although neither of these was an albino, attendant scientists are hoping that one day Snowflake will sire a youngster of a similar colour to perpetuate the genetic mutation that makes the father such a prized possession.

ABOVE RIGHT
With shoulders twice the size of those of its close relative the Chimpanzee, a male Gorilla is the most formidable of all primates. Its massive lightly-haired crown supports muscles which are closely associated with the working of powerful jaws to chew up vegetation. This feature is far more strongly pronounced in males than in females and is a useful means of identification when only the head is visible in dense mountain vegetation.

LEFT

Foraging among the greenery, a troop of Mountain Gorillas have discovered a choice meal of wild bananas in the Kahuzi-Biega National Park in Zaire. Full-grown Gorillas eat a vast amount, and as a troop will contain on average between five and fifteen individuals of all ages, it is not too surprising to find that they spend a large proportion of the day searching for fruit, bamboo and wild celery, all of which form part of their natural diet. Usually there is one mature silver-backed male in charge of the welfare of up to four adult females and perhaps ten young of various ages. The younger males are black-backed and eventually they will challenge their silver-backed patriarch for his prized position in the troop. If he is forced to step down, the once-reigning king will be driven out to lead a solitary life in the jungle.

INDEX

Numbers in italics denote illustrations

ACKNOWLEDGMENTS

The publishers would like to thank the following individuals and organisations for their kind permission to reproduce the photographs in this book:

Anthro Photo File (D J Chivers) 44, 78, 79 right, (Nancy DeVore) 87 above left, (Wrangham) 7, endpapers; Ardea (J R Beames) 66-67, (A & E Bomford) 26-27, 30, 36, 38-39, 39, (J P Ferrero) 72 left, (K Fink) 50 above, 64 left, 74-75, 81, 93, (S Gooders) 57, (M E J Gore) 91 left, (C Hagner) 62-63, (J W Mason) 50 below, (P Steyn) 25 above right, (R Waller) 73 below; Associated Press 94 below; David Attenborough 24, 29, 36-37 below; Simon Bearder 25 above left, 25 below, 68; Beth Bergman 80, 94-95; S C Bisserot 14-15, 19, 48; Ron Boardman 17 above, 32 above; Bruce Coleman 2-3, (H Albrecht) 90 below, 91 right, (Lee Lyon) 92 below, 94 below, 95, (D A J Mobbs) 88; F Erize 43 below, 46, 52 left; Jacana (Y A Bertrand) 8, (V Renaud) 90 above, (Revez CNRS) 70; Claire Leimbach 21, 65, 77, 85, 87 above right and below; J R MacKinnon 22-23, 86; R D Martin 34-35, 36-37 above; P Morris 11, 18, 32-33 below, 69 below, 73 above, 82-83; NHPA (A M Anderson) 70-71, (I Polunin) 61, (E H Rao) 69 above left; Natural Science Photos (Dan Freeman) 69 above right; Pollock 28; San Diego Zoo 53; A. Walker 17 below, 20; Shin Yoshino 1, 4-5, 13, 31, 33 above, 40-41, 47, 48-49, 51, 52 right, 54-55, 58 above and below, 59, 64 right, 72 right, 79 left; ZEFA (B Leidman) 56, (Lummer) 43 above, (Starfoto) 60-61, (A Thau) 45; Zoological Society of London 92 above.